Post–Covid Public Health: Are We Almost Seeing The Finish Line?

Post–Covid Public Health: Are We Almost Seeing The Finish Line?

Authors:

Austin Mardon

Ezzah Inayat

Tristan Ramsubag

Kanish Baskaran

Daniel Gurin

Jessica Henschel

Lea Touliopoulos

Massa Mohamed Ali

Annilea Purser

Editors:

Catherine Mardon

Huma Inayat

Golden Meteorite Press

103 11919 82 St NW

Edmonton, AB T5B 2W3

www.goldenmeteoritepress.com

ISBN: 978-1-77369-904-2

EBook ISBN: 978-1-77369-905-9

GM
PRESS

Typeset and Cover Design by Nadia Feller

Table of Contents

Turning the world upside down: What happened during the COVID–19 pandemic?

By Ezzah Inayat

In December of the year 2019, a cluster of patients in Wuhan, China began exhibiting pneumonia-like symptoms, and were unresponsive to standard treatments (Centers for Disease Control and Prevention, n.d). Over the next few weeks, the city of Wuhan began to see more cases of this mysterious illness, leaving doctors and scientists baffled at what could be the cause.Infected individuals all seemed to exhibit similar symptoms, including coughing, fever, dyspnea, and difficulty breathing, among other frightening symptoms. By January 5th, 2020, Chinese public health officials shared the genetic sequence of this new virus with the world through an online database in an attempt to find some answers. This prompted the Centers for Disease Control and Prevention to initiate an investigation of this unusual illness. By January 7th, 2020,public health officials in China were able to identify the cause of this illness to be a novel coronavirus (Centers for Disease Control and Prevention, n.d.). This virus was later named SARS-CoV-2, and scientists coined the term COVID-19 to describe the illness caused by this virus. Despite efforts to keep the virus contained, its highly contagious nature led to its spread across the globe. What started as a small outbreak in the city of Wuhan, turned into a global pandemic that has left its mark on history. This chapter will discuss the timeline of events leading up to and throughout the COVID-19 pandemic.

Some topics that will be touched on include the spread of the virus across the globe, global response to this rapid spread, advancements in treatments, as well as implications of the pandemic for the general population in terms of social, economic, and health impacts.

Spread Across the Globe

On January 13th, 2020, the Thailand Ministry of Public Health reported its first case of COVID-19. This was the first reported case of the virus outside of China. Just a week later, the virus had reportedly spread to Japan and South Korea as well (Centers for Disease Control and Prevention, n.d.). At this point, surrounding countries remained vigilant, determined to prevent the spread of the virus. The Centers for Disease Control and Prevention began to conduct screenings for passengers on connecting flights from Wuhan to airports across the United States. Passengers were screened for symptoms such as fever and cough. Despite these measures, on January 20th, 2020, the United States saw its first case of COVID-19. At this point, scientists had confirmed that the virus was airborne, spreading through human-to-human contact (Centers for Disease Control and Prevention, n.d.). This news was met with a mixture of emotions from across the globe, including fear, anxiety, and stress. At this point, there were still so many unknowns regarding the virus. Where did it come from? How did it spread? What about potential cures? Was it fatal? What was the incubation period? These questions had doctors and scientists puzzled, working hastily to find answers to these questions. The only problem was, finding these answers required time, a luxury that scientists did not have. As they worked vigorously to study SARS-CoV-2, the virus continued to wreak havoc across the globe, spreading like wildfire. By March 2020, the World Health Organization had declared a global pandemic, the first one since the 2009 H1N1 pandemic (Centers for Disease Control and Prevention, n.d.). By December

2020, just a year after Chinese public health officials identified the first cluster of patients with the virus, over 60 million cases of COVID-19 had been reported. Today, this number is close to 700 million, and is only expected to rise.

Global Response

As mentioned previously, the COVID-19 pandemic was the first pandemic in over a decade. The highly contagious nature of the SARS-CoV-2 virus is highly similar to the influenza virus, which also led to the declaration of a global pandemic over a century ago in 1918. Much has changed since then. Technological innovation has led to the advancement of health care. Research has also allowed scientists to gain insight on various viruses, allowing for quicker and more effective pharmaceutical drug development. Given this information, it is important to consider the following; how did the world handle the COVID-19 pandemic compared to prior pandemics? For one, globalization (although beneficial as a whole) made it very easy for the virus to spread across the globe, and as a result, countries across the world had to work together in a combined effort to mitigate the spread of COVID-19. Epidemiologists from various countries worked together to study the virus's mode of transmission and pattern of spread, allowing public health officials to use this information to create effective mitigation strategies. Many countries took a strict lockdown approach towards the pandemic. In late January, China implemented a strict lockdown which lasted approximately two months (Centers for Disease Control and Prevention, n.d.). The streets were eerily quiet, empty with the exception of health and security professionals, as well as emergency vehicles. In March, Italy took a similar approach in response to the rapid spread of the virus in vulnerable populations. Other countries, such as Canada and the United States, also attempted to mitigate the spread of COVID-19 using lockdown measures. Schools and universities closed down, and we saw

9

a transition to virtual work and learning. Teachers, students, and employers had to adapt to changing work conditions, and had to overcome various barriers to communication in virtual environments. Mask mandates were introduced across the globe after many studies proved the efficacy of masking at preventing the spread of COVID-19. The use of social distancing, masking, and proper hygiene practices were relied on while scientists worked on developing a vaccine that was effective against the virus.

Advancements: Vaccines, At–Home Testing, Symptom Management

Over the course of the year 2020, life was essentially put on hold as the world attempted to cope with COVID-19. Healthcare systems were overwhelmed with patients, schools and workplaces had shut down, social distancing measures were strictly implemented, all while scientists searched for answers on potential solutions to this pandemic. Although mitigation efforts proved effective, these efforts were not suitable for the long-term. Ultimately, scientists were searching for a cure, in this case, a vaccine that could potentially reduce the spread of the virus as well as reduce its impacts. This would lead to improved patient outcomes, protecting vulnerable populations while allowing life to "return to normal". Unfortunately developing an effective COVID-19 vaccine would take time, and required the general public to remain patient while continuing to adhere to social distancing measures.

The race to develop an effective vaccine was quite competitive. Companies all over the world were working tirelessly to be the first to put out a vaccine, as they knew this would come with financial and political incentives. Companies leading the race included Pfizer, which had partnered with pharmaceutical company BioNTech to develop a vaccine that demonstrated 95% efficacy in phase 3 clinical trials (Katella, 2022). At the same time, Moderna was working to develop a vaccine with similar efficacy. Both Pfizer and Moderna relied on mRNA technology to develop vaccines that were highly effective. This type of technology differs from traditional vaccine development in the sense that mRNA vaccines require laboratory generated mRNA targeted towards a specific protein to be injected into the body. In the case of Pfizer and Moderna, this mRNA was targeted towards the SARS-CoV-2 spike protein. Injection of this mRNA into the body via vaccination allows the body to create this spike protein, which in turn triggers an immune response that will produce antibodies. These antibodies help to prevent infection in the future (Centers for Disease Control and prevention, 2022). Both vaccines created by Pfizer and Moderna were authorized for use in those above 16 and 18 respectively, in early 2021. Vulnerable populations, such as essential frontline workers and those over the age of 70 were prioritized for vaccine administration as doses were limited. Slowly, eligibility expanded to the general public. Although Pfizer and Moderna proved to be the most effective vaccines in clinical trials, other vaccines were in use across the globe. Astrazeneca, a traditional vaccine that did not rely on mRNA technology, was also administered in many countries despite its lower efficacy. Chapter 5 will further discuss the vaccine development process.

While vaccine development was underway, the world also saw the innovation of at-home rapid tests. At the beginning of the COVID-19 pandemic, in order to get tested for the virus, symptomatic individuals would have to visit a testing facility for a Polymerase Chain Reaction (PCR) test. The wait times at these testing facilities were often lengthy, and results typically took a few days. This was inconvenient, as by the time test results were received, symptomatic individuals may have risked spreading the virus to others. In addition, there was an increased risk of transmission at these testing facilities. With the development of at-home antigen tests,individuals could test for the virus with a simple nasal swab, receiving results within five minutes. This highly innovative testing method reduced wait times for PCR tests, and allowed symptomatic individuals to receive quicker test results.

The Battle Rages On: Variants and Waning Immunity

When word of an effective COVID-19 vaccine spread, people across the globe were overjoyed. Everyone began to imagine a return to normal life. After months of not being able to attend school or work, travel, or spend time with loved ones, people were eager to return to their normal daily activities. Things gradually began to change, with businesses increasing their working hours, and limits on gatherings slowly being decreased. The end of the pandemic seemed to be in sight, that is until the harsh realities of waning immunity and virus variants struck.

The world began to see emerging variants of the SARS-CoV-2 virus, including the Delta and Omicron variants. These variants proved to be better at evading immunity as a result of the vaccine, and also proved to be highly transmissible compared to the original virus. Chapter 2 will further investigate these variants in detail. Scientists also posed the concern of waning immunity, as vaccine protection eventually fades. With the emergence of new virus variants, and waning immunity with time after vaccination, the world began to see a rise in COVID-19 cases once again. This was extremely disheartening to many, and left people wondering if the pandemic would ever truly end.

Implications of the Pandemic: Social, Economic, and Health Impacts

As the pandemic raged on, life completely changed for everyone as the world tried to cope with COVID-19. As mentioned earlier on in this chapter, many public health strategies were introduced to mitigate the spread of COVID-19 across the globe, including lockdown measures and social distancing. Although these measures were relatively successful at preventing the spread of COVID-19 until an effective vaccine became available, the pandemic had significant implications on other aspects of life.

In terms of social impacts, the pandemic had significant impacts on social interaction and mental health. According to the World Health Organization (WHO), there was a 25% increase in the prevalence of anxiety and depression worldwide throughout the first year of the COVID-19 pandemic (World Health Organization, 2022). There are many potential reasons for this. Throughout the pandemic, the world saw a transition to virtual communication. Platforms such as Microsoft Teams and Zoom dominated the corporate and educational world. These platforms allowed for employees and students to remain productive without having to physically interact with others and risk spreading the virus. Although these platforms allow for productivity, virtual communication and in-person interaction are very different things. Navigating barriers to communication in such environments can be difficult, and as a result, left many people feeling a sense of disconnect throughout the pandemic. This may have led to feelings of social isolation, possibly contributing to increased depression and anxiety levels. Other measures may have also contributed to feelings of social isolation. For instance, caps on large gatherings such as weddings and funerals led to further social disconnect from loved ones. Restricted visitors in hospitals and nursing homes also contributed to this. In addition, social distancing and work-from-home measures may have

13

forced individuals to stay in toxic home environments, further contributing to poor mental health. Statistics show that young people in particular experienced significant declines in mental health (World Health Organization, 2022). Closures of university residences robbed students of the social aspect of learning, making the adjustment to university life particularly difficult for incoming first year students. School closures also limited crucial social interaction within children, making it difficult for them to develop the essential collaboration and social skills they typically do in school. The economic impacts of the pandemic, which will be discussed further on in this chapter, may have also contributed to poor mental health in financially struggling families. These factors, coupled with the fear and stress of contracting the virus (or having a loved one contract the virus) can explain the sharp increase in the prevalence of depression and anxiety. This increase posed a serious cause for concern, and led to many global mental health initiatives aiming to combat mental illness.

In terms of economic impacts, Canada saw the loss of millions of jobs throughout the pandemic, with the unemployment rate soaring to 13% during April of 2020 (Canadian Broadcasting Corporation, 2020). Reduced hours as part of measures to reduce the spread of COVID-19 led to many workers being fired from their jobs. Small businesses were hit particularly hard during the pandemic, as these businesses did not have the means to quickly transition to virtual means of productivity, and were hit with reduced business. Many of these small businesses were forced to temporarily (and sometimes permanently) close down, further contributing to the negative economic impacts of the pandemic. Similar impacts wereexperienced across the world. This had significant implications for the global economy. Many countries tried to provide grants and funding to help financially struggling families. For example, the Canada Emergency Response Benefit (CERB) and Canada Emergency Student Benefit (CESB) were both programs that were introduced to help financially struggling Canadians throughout the pandemic. Despite this, many Canadians experienced financial struggles, which only contributed to poor health outcomes as discussed earlier.

In terms of health impacts, we have already discussed the rising rates of mental illness, especially among young people in particular, however the pandemic has also had significant physical health impacts on society. In the midst of the COVID-19 pandemic, a phenomena commonly referred to as "Long Haul Covid" soon began to make headlines. Many patients who had recovered from COVID-19 reported experiencing long-term symptoms even after recovery. These symptoms include autoimmune disorders and multiorgan impacts affecting the lungs, kidneys, heart, and other organs in the body (Centers for Disease Control and Prevention, 2022). Some people reported never regaining their lost sense of smell or taste after recovering from COVID-19, while others reported developing tinnitus as a result of the virus. Other commonly reported long term impacts included chronic fatigue, arrhythmia, and arthritis. These long-term impacts were referred to as "Long COVID" or "Post COVID conditions" and left doctors baffled at their cause and treatment. Upon investigation, scientists concluded that COVID-19 is a mysterious illness, and can have differing impacts on different people. These impacts may take weeks, months, or even years to dissipate, which is an unfortunate reality for those with Long COVID. Further research is needed to tackle the issue of Long COVID and improve the deteriorating quality of life of those struggling with Post COVID conditions. In addition to Long COVID, the COVID-19 pandemic resulted in many people resorting to a sedentary lifestyle. With gyms closed, and social interaction limited, many people found themselves cooped up in their homes with little to no physical activity in their daily life. A sedentary lifestyle is a risk factor for obesity and diabetes, which was extremely concerning considering hospital systems were already overwhelmed with COVID-19 patients throughout the pandemic. Studies show that obesity rates increased by approximately 10% in 5-11 year olds throughout the pandemic (Beaumont, 2022). This can have future implications for the health of these individuals as they grow older. Finally, as a result of overwhelmed health care systems throughout the pandemic, many individuals struggling with other illnesses were also impacted. For instance, many elective surgeries and treatments were delayed as a result of short staffed hospitals, and hospital wait times remained at an all-time high (Canadian Institute for Health Information, 2022). This can have negative implications on patient outcomes and quality of life.

Living with the Virus, What Next?

As of December 2022, there have been a whopping 700 million cases of COVID-19 worldwide. This figure is only based on laboratory confirmed cases, so in reality, the number of people who have caught the virus is likely exponentially larger. Among many unknowns, scientists are in agreement on one thing: the virus is here to stay. As a result, the world is currently entering a new era, one where people must learn to live with COVID-19. To many, this feels strange. Learning to live with the virus that wreaked havoc on society for two years is a frightening thought, however scientists argue that we are better prepared to deal with SARS-CoV-2 now than ever before. Vaccines provide some level of immunity against the virus, and scientists are working on the creation of updated vaccines targeted towards COVID-19 variants. With constantly updated vaccines, scientists are hopeful that they can protect vulnerable populations.

The world has seen a drastic decline in new COVID-19 cases so far, however epidemiologists still expect to see seasonal waves of cases. Fortunately, we now know enough about the virus to be prepared for such waves. During these waves, scientists have advised that people limit social contact as much as possible and wear masks when they can. For individuals with symptoms of COVID-19, scientists advise that they limit social contact in order to prevent its spread. Moving into 2023, after three long and grueling years, the end of the pandemic is finally in sight. Businesses, malls, and wedding venues have opened up to full capacity, students have returned to in-person learning, and the majority of mandates have been dropped. People are still having difficulty adjusting to this new "normal", however scientists and public health officials are hopeful that the worst has passed.

References

Beaumont. (2022). Has the Pandemic Affected Obesity Rates? Retrieved December 31st, 2022, from, https://www.beaumont.org/health-wellness/blogs/has-the-pandemic-affected-obesity-rates

Canadian Broadcasting Corporation. (n.d). Canada lost nearly 2 million jobs in April amid COVID-19 crisis: Statistics Canada. Retrieved December 31st, 2022, from, https://www.cbc.ca/news/business/canada-jobs-april-1.5561001

Centers for Disease Control and Prevention. (n.d.). COVID-19 Timeline. Retrieved December 31st, 2022, from, https://www.cdc.gov/museum/timeline/covid19.html

Centers for Disease Control and Prevention. (n.d.). Understanding how COVID-19 vaccines work. Retrieved December 31st, 2022, from, https://www.cdc.gov/coronavirus/2019-ncov/vaccines/different-vaccines/how-they-work.html

Centers for Disease Control and Prevention. (2022).
Long COVID or Post-COVID Conditions. Retrieved
December 31st, 2022, from, https://www.cdc.gov/coro-
navirus/2019-ncov/long-term-effects/index.html#:~:tex-
t=Some%20people%2C%20especially%20those%20
who,kidney%2C%20skin%2C%20and%20brain.

Katella, K. (2022). Comparing the COVID-19 vaccines: how
are they different? Retrieved December 31st, 2022, from,
https://www.yalemedicine.org/news/covid-19-vaccine-com-
parison#:~:text=How%20well%20it%20works%3A%20
When,independent%20analysis%20by%20the%20FDA.

World Health Organization. (2022). COVID-19 pandemic triggers 25% increase in prevalence of anxiety and depression worldwide. Retrieved December 31st, 2022, from,
https://www.who.int/news/item/02-03-2022-covid-19-pan-
demic-triggers-25-increase-in-prevalence-of-anxiety-and-
depression-worldwide

Understanding the Covid–19 Variants

By Lea Touliopoulos

When the first news of COVID-19 began to sweep through the world at the beginning of 2020, many people assumed that COVID would be a short-term problem and would be resolved in a couple of months. However, almost three years later COVID is still proving to be present, even after the development of vaccines, many research advancements, and many public health measures put into place with the goal of slowing the disease. How is it possible for a virus to continue to infect and re-infect so many people over several years? Why have the symptoms and severity of COVID-19 shifted over the years that it has been present? While these are complicated questions with complicated answers, one of the reasons that COVID-19 has continued to be such a persistent problem over the past two years is that it has to continue to evolve leading to the appearance of a multitude of new variants. These new variants have been able to spread globally and are largely responsible for the continued circulation of COVID-19 within communities.

In order to understand the journey of COVID-19 and the appearance of new variants, it is important to have a basic understanding of what COVID-19 is. COVID-19 is a virus that belongs to a class of viruses called coronaviruses (Mohamadian et al., 2021). It has been given the name SARS-CoV-2 and is closely related to other coronaviruses such as SARS-CoV (otherwise known as the Severe Acute Respiratory Syndrome-coronavirus) which was responsible for the outbreak in 2002 and 2003, as well as the MERS-CoV (Middle East Respiratory Syndrome-coronavirus) which was responsible for the outbreak in 2012 (Mohamadian et al., 2021). These three coronaviruses are part of the seven human coronaviruses that have been discovered so far (Mohamadian et al., 2021). The reason why SARS-CoV and MERS-CoV are more well known is that they both also caused epidemics, with SARS-CoV having a mortality rate of approximately 9.5% and MERS-CoV having a mortality rate of approximately 34.4% (Mohamadian et al., 2021). While SARS-CoV-2 (colloquially known as COVID-19) does appear to have a lower mortality rate than both SARS-CoV and MERS-CoV, it does appear to have higher transmissibility and more varied clinical manifestations, which played a role in it's rapid spread across the globe (Mohamadian et al., 2021). This resulted in the large number of people contracting COVID-19 and becoming ill due to the virus.

Diving into the structure of SARS-CoV-2, it is a positive-sense single-stranded RNA with a 5'cap, 3'UTR poly(A) tail (Mohamadian et al., 2021). Its genome consists of 14 open reading frames, which encode non-structural proteins for virus replication, along with structure proteins such as the spike (S) protein, the envelope (E) protein, the membrane (M) protein, and the nucleocapsid (N) protein (Mohamadian et al., 2021). One key element of the SARS-CoV-2 virus is that it is able to bind to angiotensin-converting enzyme 2 receptors, otherwise known as ACE2 receptors (Mohamadian et al., 2021). ACE2 receptors are found in many different tissues in humans, including the lungs, kidneys, gastrointestinal tract, heart, liver and blood vessels. The virus is able to do this through its S protein, which is a transmembrane protein that is composed of two units, one for receptor binding, otherwise known as the S1 subunit, and one for cell membrane fusion, otherwise known as the S2 subunit (Mohamadian et al., 2021).

It is thought that the SARS-CoV-2 originated from the bat coronavirus (BatCoV), as it has been shown through genome sequencing that SARS-CoV-2's genome is 96% identical to the bat coronavirus (Mohamadian et al., 2021). It is believed that the outbreak began in a seafood and poultry market in Wuhan China (Mohamadian et al., 2021). Some interesting background information about Wuhan is that it is a city of 11 million people located in central China. Similar to SARS-CoV and MERS-CoV, before SARS-CoV-2 was able to have human to human transmission, it had a zoonotic source. A zoonotic source is referring to the fact that the virus came from an animal prior to mutating in order to be able to be transmitted amongst humans.

With respect to the SARS-CoV-2 virus, the zoonotic source is likely bats (Mohamadian et al., 2021). SARS-CoV-2 is the virus that first was present in the early months of 2020 (Taylor, 2021). The timeline of SARS-CoV-2 is that on December 21, 2019 the WHO China Country Office was first informed of cases of pneumonia in Wuhan (Taylor, 2021). However, this pneumonia was different from other more common types of pneumonia and had an unknown etiology. What we now know to be SARS-CoV-2 was soon seen to be very infectious, and quickly spread around the world, causing the WHO to declare COVID-19, otherwise known as SARS-CoV-2, as a global health emergency on January 30, 2020 (Taylor, 2021). While many people rightly viewed this as alarming news, the public sentiment was that pandemic would only last for a short time period before resolving itself, especially after the creation of multiple vaccines. However, the COVID-19 pandemic has surpassed many peoples' original predictions of only being relevant for several months, and is still circulating almost three years later. One question many people may be asking themselves is how COVID-19 is still managing to infect so many people, especially those who have been vaccinated and those who have already been infected and recovered from COVID-19.

While the answer to these questions are multifaceted, one factor that has been aiding COVID-19 to continue to actively circulate amongst vaccinated populations who should already have antibodies to the virus is that it has been able to mutate into many different variants. Variants are when a virus mutates as it replicates, meaning that its viral genetic sequence will differ from what it originally was (Shah, 2021). If the virus differs enough from the original virus, it will often be labelled as a new variant of the original virus (Shah, 2021). This means that the variant will share many of the features with the original virus, but will also behave in a way that is different enough that the original virus and the new variant can be differentiated from each other (Shah, 2021). Some of the mutations that occur can give the variant a selective advantage compared to the original virus (Shah, 2021). This can allow the variant to be able to successfully reproduce, meaning that it will become more common (Shah, 2021). This new variant can be more or less dangerous than the original, more or less infectious, or have other distinct traits compared to the original virus depending on the random mutations that have occurred (Shah, 2021).

Returning back to the discussion about the COVID-19 virus and its variants, one thing that is important to understand is that the virus is constantly changing. There have been many small mutations in its genetic composition that have created different variants, but not all of them have been significantly different from the original virus that they have been noteworthy. However, there have been many noteworthy variants since COVID-19 emerged at the very end of 2019. Some of these more well known variants are the Alpha variant, the Beta variant, the Delta variant, and more recently the Omicron variant. As is somewhat evident from the variants' names, the World Health Organization names the variants using letters from the Greek alphabet. This means that the first variant was named the Alpha variant, the second the Beta variant, and then the naming continued from there. One positive of this naming system is that it prevents people from referring to variants by the name of the country where the variant was first detected (Abu-Raddad et al., 2021).

Starting off with the Alpha variant, which was first noted in the United Kingdom during September 2020 (Abu-Raddad et al., 2021). One reason why the Alpha variant was able to become the dominant variant in many different places around the globe is that it was 50% more transmissible than earlier strains (Abu-Raddad et al., 2021). The Alpha variant was very prevalent in Canada throughout the first months of 2021, and was thought to be a major driver of Canada's third wave of COVID-19 cases during that time period (Abu-Raddad et al., 2021). Some symptoms of the COVID-19 Alpha variant are fever or chills, a cough, shortness of breath, difficulty breathing, fatigue, muscle and body aches, headaches, a loss of the senses of taste and/or smell, a sore throat, congestion, a runny nose, nausea, vomiting, and diarrhea (Patterson, 2021). Those who were infected by the Alpha variant of COVID-19 would not necessarily present with all of those symptoms, but most would present with at least several from that list.

Another variant of the COVID-19 virus is the Gamma variant. This variant was first found in Brazil, early during December 2020 (Public Health Ontario, 2021). One reason that the existence of this variant was considered to be concerning was because preliminary data suggested that it had both an increased transmissibility and an increased severity compared to other variants (Public Health Ontario, 2021). However, the Gamma variant was less successful at spreading across the globe compared to many other variants; its global prevalence remained relatively low, hovering at around 2% between the months of December 2020 and July 2021 (Public Health Ontario, 2021). However, the Gamma variant was able to become the dominant strain in some regions of Brazil, which was worrisome as there was epidemiological evidence from a variety of studies that showed that the Gamma variant had an increased risk of both hospitalisation and ICU admissions compared to other variants (Public Health Ontario, 2021).

Moving on, one of the first more well known variants to spread to North America was the Delta variant. The Delta variant was first seen in India, in the late months of 2020, but it was able to quickly spread through the world (Katella, 2021). It eventually became the most discussed, as well as the most well known strand of COVID-19 for most of 2021, before a different variant was able to take over later in 2021.

One of the reasons that the Delta variant was able to become the dominant variant is because it was more than two times as infections as previous viruses (Katella, 2021). Coming at a time when many people had already been infected by COVID-19 or had already had a complete vaccination series, COVID-19 needed to become more infectious in order to keep being transmitted (Katella, 2021). Unfortunately Delta was able to succeed in this. An example of this comes from the state of Connecticut. There the Delta variant was estimated to be 80% to 90% more transmissible than the previous alpha variant (Katella, 2021). This means that 80% to 90% more people were getting infected by COVID-19 (Katella, 2021). This would explain a lot of the second surge of COVID-19 in June 2021, at a time when the number of people sick with COVID had been decreasing prior to the spread of the Delta variant. Even those who had been vaccinated were susceptible to getting infected with the Delta variant, which is why booster shots began to be encouraged at that time, with the goal of preventing the spread of the Delta variant of COVID-19 (Katella, 2021).

Now that the transmissibility of the Delta variant has been discussed, let's turn to the severity of the disease that it could cause. The Delta variant unfortunately caused a much more severe disease than the original COVID-19 strain (Katella, 2021). This was especially evident in those who had not yet been vaccinated. This meant that those who were getting infected with the Delta variant of COVID-19, and who were not yet vaccinated, were getting hospitalized at higher rates than when the alpha variant of COVID-19 was the dominant strain (Katella, 2021).

Moving on to the symptomatology of the Delta variant of COVID-19, some common symptoms that those infected with the Delta variant might present with are a headache, a sore throat, a runny nose and a fever (Patterson, 2021). Some differentiating factors between the symptoms of the Delta variant and other variants are that a persistent cough, as well as the loss of the sense of smell and taste are not as likely to occur with the Delta variant (Patterson, 2021).

However, one positive thing concerning the Delta variant of COVID-19 was that COVID-19 vaccines have been found to protect people against the Delta variant (Katella, 2021). This means that those who have been vaccinated are much less likely to be hospitalized, to have severe long lasting illness, or to die if they become infected with the Delta variant of COVID-19 (Katella, 2021). However, it is important to keep in mind that no vaccine is 100% effective and that vaccinated people can still become infected and spread the virus to unvaccinated people, as well as vaccinated people.

This is one of the reasons that there were still many public health protocols in place during the time the Delta variant of COVID-19 was the dominant variant. (Katella, 2021) Public health measures such as advising people to stay home when they're sick, enforcing social distancing, promoting hand hygiene, and encouraging people to maintain good overall health were still effective ways to prevent the spread of viruses, including the Delta variant of COVID-19 (Katella, 2021).

While the Delta variant was the dominant strand in most of North America for the majority of the summer of 2021, the Omicron variant of COVID-19 eventually took over as the dominant strand of COVID-19 (CDC, 2021). The Omicron variant was first detected near the end of 2021, in November (Kwon, 2022). It was first detected in South Africa (though retrospective data showed the existence of earlier cases in the Netherlands (Katella, 2021)), but it was able to spread around the world faster than any other of the previous COVID-19 variants (Kwon, 2022). Furthermore, the COVID-19 vaccines seemed to be less effective against the Omicron variant, as many people who had received complete vaccination series and even boosters were still getting infected with COVID-19 and were symptomatic (Kwon, 2022).

Before discussing the symptomatology of the Omicron variant it is important to understand how its molecular structure differs from other COVID-19 variants. The Omicron variant is thought to be one of the variants with the most structural differences (Kwon, 2022). With over 30 mutations in the spike (S) protein that is found on the surface of the virus, the Omicron variant is able to easily latch on to the host cells and therefore is able to infect those cells (Kwon, 2022). Even more significant is the fact that 15 of those mutations are found in the receptor binding domain (RBD) of the S protein (Kwon, 2022). The RBD is the area of the S protein that binds to the ACE2 receptors and allows the virus to enter a person's cells (Kwon, 2022). Another area of importance in the S protein is the N-terminal domain, and the Omicaron variant has 11 mutations in this region as well (Kwon, 2022). The reason that these mutations are so significant is that they have changed the areas of the S protein that are recognized by antibodies (Kwon, 2022). Recall that antibodies are what the human body uses to fight viruses, and are the underlying principle of how vaccines provide immunity. Since the Omicron variant has these mutations, antibodies from previous COVID-19 infections or from vaccines will not be nearly as effective at recognizing the virus, which means that they will not be able to protect against the Omicron variant of COVID-19 as well as they could with prior variants (Kwon, 2022). This is why the Omicron variant was able to infect so many of those who have received COVID-19 vaccines or who have already recovered from a COVID-19 infection (Kwon, 2022).

Now that the molecular structure of the Omicron variant has been discussed, how does the symptomatology of the Omicron variant differ from the symptomatology of other circulating variants? Overall, there is not a large difference in symptoms in those infected with the Omicron variant compared to those infected with other variants (CDC, 2021). Luckily, data had shown that the Omicron variant does cause less severe illness and less death in general compared to prior variants (CDC, 2021). A common symptom from infection of the Omicron variant is severe fatigue, and it appears that common symptoms from other variants, such as the loss of the sense of taste and the sense of smell, are much less common from infection of the Omicron variant (CDC, 2021). However, due to the Omicron variant spreading more easily and antibodies from vaccines and past infections being much less effective than on past variants, there was still a surge in cases that threatened to overwhelm many health care systems (CDC, 2021). This was a major contributor to the increase of public health measures across North America from December 2021 to January 2022, as overwhelmed health care systems make it much more difficult for healthcare professionals to treat severe cases, as well as to treat other medical emergencies that will occur that are unrelated to COVID-19.

Throughout the beginning of 2022, the BA.5 strain of the Omicron variant was the dominant strain of COVID-19 throughout much of the United States (Kwon, 2022). However, as viruses are constantly evolving and mutating, it is unsurprising that two different sub variants of the Omicron variant have since become the dominant strains. Around the middle of November

2022, the subvariants BQ.1 and BQ.1.1 have since become the dominant strains in the United States of America (Kwon, 2022). One reason that this is noteworthy is because the BQ.1 strain and the BQ.1.1 are both thought to be less susceptible to the antibodies from vaccines than the prior BA.5 strain (Kwon, 2022). This means that it is less likely that people will have immunity to the BQ.1 and the BQ.1.1 strains of the Omicron variant even if they have been vaccinated, received multiple boosters and/or have recovered from a previous COVID-19 infection (Kwon, 2022).

Evidently, since COVID-19 first appeared at the end of 2019 in China there have been a vast variety of different variants and strains that have circulated the globe. Some of the major variants in North America that became dominant strains were the Delta variant, as well as the Omicron variant. However, what many people may be asking themselves is what will the last COVID-19 variant be? Unfortunately, this question may be unanswerable, as variants are always part of the natural progression of viruses (Kwon, 2022). There will always be random mutations giving viruses the ability to evolve. As long as there is a COVID-19 outbreak somewhere, there will always be the likelihood that a new variant will evolve (Kwon, 2022). Ideally, the virus will evolve to become less and less severe, and maybe even vanish at some point in the future, but up to that point, there will always be new variants of COVID-19 emerging and circulating (Kwon, 2022).

Citations

Abu-Raddad, L. J., Chemaitelly, H., & Butt, A. A. (2021). Effectiveness of the BNT162b2 Covid-19 Vaccine against the B.1.1.7 and B.1.351 Variants. New England Journal of Medicine, 385(2), 187–189. https://doi.org/10.1056/NE-JMc2104974

CDC. (2020, February 11). Coronavirus Disease 2019 (COVID-19). Centers for Disease Control and Prevention. https://www.cdc.gov/coronavirus/2019-ncov/variants/index.html

Katella, K. (2022, March 1) 5 Things To Know About the Delta Variant. Yale Medicine. Retrieved December 31, 2022, from https://www.yalemedicine.org/news/5-things-to-know-delta-variant-covid

Kwon, D. (2022). Omicron's molecular structure could help explain its global takeover. Nature, 602(7897), 373–374. https://doi.org/10.1038/d41586-022-00292-3

Mohamadian, M., Chiti, H., Shoghli, A., Biglari, S., Parsamanesh, N., & Esmaeilzadeh, A. (2021). COVID-19: Virology, biology and novel laboratory diagnosis. The Journal of Gene Medicine, 23(2), e3303. https://doi.org/10.1002/jgm.3303

Patterson, K. (2021, August 12). Alpha Variant vs. Delta Variant – How Are the Symptoms Different? Franciscan Missionaries of Our Lady Health System. https://health.fmolhs.org/body/covid-19/alpha-variant-vs-delta-variant-how-are-the-symptoms-different/

Public Health Ontario. (2021, July 22) COVID-19 Gamma Variant: Risk Analysis and Implications for Practice.

Shah, D. (2021) What is a variant? An expert explains I News. (n.d.). Wellcome. Retrieved December 31, 2022, from https://wellcome.org/news/what-variant-expert-explains

Taylor, D. B. (2021, March 17). A Timeline of the Coronavirus Pandemic. The New YorkTimes. https://www.nytimes.com/article/coronavirus-timeline.html

The Crash of the Healthcare System: ER/ICU, Wards, and Beyond

By Daniel Gurin

Introduction

Serious disruptions to worldwide healthcare systems were felt throughout the world during the COVID-19 pandemic. Nations faced a strong disbalance between the demand and supply of medical resources when they were needed most. Despite the pandemic being in its ending stages, many healthcare systems have yet to recover and may take a while to do so. This chapter will explore some of the factors that contributed to the crash of healthcare systems during the pandemic, as well as some downstream effects on other facets of healthcare and potential solutions going forward.

Healthcare Worker Burnout

Caused by insufficient support, ever-increasing workloads, underfunding of public healthcare infrastructure and the accumulating moral distress of being unable to provide the care patients require, the COVID-19 pandemic exacerbated the burnout faced by healthcare workers. Hundreds of thousands of healthcare workers were pushed to their limit, with 52% of U.S. nurses and 20% of U.S. doctors stating that they plan on leaving their clinical practices after dealing with the pandemic, according to the Mayo Clinic Proceedings and American Nurses Foundation (Murthy et al., 2022). Despite burnout being a major strain on healthcare during the pandemic, the issue's full effects may not have yet been felt in full. The U.S. Bureau of Labor Statistics recently projected a nurse shortage of 1 million by the end of 2022, caused in part by the burnout experienced by existing nurses and the aversion to joining the healthcare workforce from potential new nurses (Murthy et al., 2022). Canada's future seems to be heading in a similar direction, with a predicted nurse shortage of 117,600 by 2030. On a global scale, the International Council of Nurses estimated that an additional 13 million nurses will be needed to fill the global nurse shortage (ICN, 2022). Seeing this, it is clear that healthcare worker burnout is a threat to the national health and economic security of many countries.

Added to the physical factors that have driven millions of nurses to their brink are the anger and frustration from patients and communities. At the beginning of the pandemic, nurses were praised as heroes and thousands gathered outside waving flags and posting "thank you" signs on their yards. Later into the pandemic, feelings of hostility started overpowering the gratitude, with patients and communities getting angry at long wait times, deferred surgeries, and the loss of trust in the healthcare system (Murthy et al., 2022). Healthcare professionals were also often tasked with making challenging decisions and were constantly criticized for their solutions.

However, healthcare worker burnout did not begin as a result of COVID-19. Even before the pandemic, healthcare professionals were undergoing immense levels of stress and burnout. Canadian Medical Association (CMA) conducted a study in 2019 which found that about 19% of Canadian physicians went to work in 2018 at least five times despite feeling either physically ill or mentally distressed (McEvoy & Thompson, 2022). In the same year, 53% of the study's respondents stated that they were dissatisfied with the resources and management of their workplace (McEvoy & Thompson, 2022). Therefore, healthcare worker burnout is far from a new concept, and the pandemic only magnified its factors.

Financial and Economic Burdens

Added to the death toll and long-term health damage caused by the COVID-19 pandemic, the financial and economic toll is also quite strong. The total financial cost from COVID-19-related measures was over $534 billion in the U.K., $222.5 billion in the U.S., and $576 billion in Canada (Richards et al., 2022). These figures include extra funding to healthcare systems, financial aid to businesses, resources needed for non-pharmaceutical interventions (NPIs), and other measures. The pandemic was a direct cause of GDP losses, as most countries experienced a loss in productivity due to lockdowns and restrictions (Richards et al., 2022).

Data from the Canadian Institute for Health Information (CIHI) found that total Canadian health spending increased by over 12% between 2019 and 2020 – triple the 4% rate seen between 2015 and 2019 (O'Toole, 2021). This dramatic increase paired with a contracted economy during the pandemic meant healthcare spending quickly outpaced the growth of the Canadian economy. Historically, Canada has spent roughly 40% of total provincial and territorial budgets on healthcare. Of this spending, hospitals account for roughly a quarter of the costs, followed by drugs and physicians at 14% and 13%, respectively (O'Toole, 2021). Seniors and other geriatric patients are the costliest age group in terms of healthcare budgets, accounting for about 45% of total public health funds. In 2019 alone, roughly $78 billion was spent on seniors (O'Toole, 2021). Overall, the pandemic caused the single largest spike in healthcare spending ever recorded in Canada, and many places around the world as well.

On a patient level, those diagnosed with more severe COVID-19 were associated with higher costs. The factors that drove these higher costs included ICU admissions and hospital resources such as mechanical ventilators. Consistently among multiple countries, the costs for patients admitted into an ICU were much higher than those that didn't require intensive care. For those that did require ICU admission, mechanical ventilators accounted for an extra $2,840.06 ± 470.52 to $4078.42 ± 744.55 per patient (Richards et al., 2022).

Furthermore, hospitals are seeing decreases in revenue figures as a direct result of labour shortages. A report run by Kaufman Hall found that two-thirds of respondents stated that their facilities were running at less than full capacity due to staff shortages (Robinson et al., 2022). Hospitals and other medical facilities have had to reach out to external contract labour in order to fill gaps during parts of the pandemic. Multitudes of respondents in the report have also said that their organizations' plans of transitioning back to a stable labour force had timelines of 3-4 years, meaning staff shortages and waiting times will likely persist for at least the near future (Robinson et al., 2022).

Shortage of ICU Beds & Lifesaving Medical Equipment

The COVID-19 pandemic saw a massive surge in demand for lifesaving medical measures. The Canadian Institute for Health Informatics (CIHI) conducted a report in 2021 that found a monthly increase of roughly 3,000 inpatient admissions between March 2020 and June 2021 (The Royal Society of Canada, 2022). It also found a cumulative 14,000 additional ICU patients in that timeframe compared to pre-pandemic levels. Prior to the pandemic, common respiratory illnesses that required the same level of treatment included pneumonia and chronic obstructive pulmonary disease (The Royal Society of Canada, 2022). COVID-19-related admissions quickly outpaced all other causes, and by the spring of 2021 (Wave 3), the demand for ICU care and ventilators had increased by about 400% compared to pre-pandemic levels (The Royal Society of Canada, 2022).

Each wave of COVID-19 brought slightly different scenarios upon hospitals. The third wave was arguably the most severe and resulted in the most ICU admissions across Canada. Wave 3 and 4 both involved the delta variant of the coronavirus, but the fourth wave saw slightly fewer ICU admissions overall (The Royal Society of Canada, 2022). The fifth wave, caused by the omicron variant, saw the highest number of cases nationally, but because of the variant's relatively less dramatic effects, resulted in far fewer ICU admissions (The Royal Society of Canada, 2022).

Canada's shortage of ICU beds and other acute medical facilities during the pandemic is a complex issue. Hospitals and provincial health authorities have put in effort in recent years before the pandemic to lean out acute care to make it as efficient as possible. In doing so, they reduced the number of acute care hospital beds and allocated facilities elsewhere (The Royal Society of Canada, 2022). As it stands, Canada has among the lowest number of ICU beds in the Organization for Economic Co-operation and Development (OECD). Because of this, Canada saw extremely high acute care bed occupancy throughout the pandemic, meaning many decisions had to be made regarding who gets appropriate care and when (The Royal Society of Canada, 2022). An average occupancy of 85% is judged to be a safe level in reducing the risk of bed shortages, but Canada has long been outside of this safety margin. Even before the pandemic, the majority of Canada's large ICU facilities were operating near the 100% occupancy level (The Royal Society of Canada, 2022). When the pandemic struck, many patients who relied on ICUs and other acute care were displaced to accommodate for acute cases of COVID-19. Based on a Government of Canada report, this contributed to a backlog of over 700,000 surgeries as of 2022. ICU occupancy levels exceeding 80% have been found to correspond with increased acute care mortality as well as ICU readmission within 7 days of discharge (The Royal Society of Canada, 2022). Countries with higher numbers of ICUs available were able to fair much better through the pandemic. Germany, which has led in terms of ICU beds by population, was even able to accommodate patients in neighbouring European countries whose ICU beds filled up early in the pandemic (The Royal Society of Canada, 2022).

More on Surgical and Other Procedural Backlogs

A study by Massachusetts General Hospital found that many forms of surgical procedures deferred due to the pandemic have yet to recover to pre-pandemic levels. The resulting backlogs have put a strain on the healthcare system and could have serious health and cost implications in the near future. The procedures that have faced the largest declines are surgical oncology, cardiac, urology, orthopedic and general surgery (Marquedant, 2022).

In Canada, traditional urgency has always been prioritized, meaning the surgeries most essential to patients' survival are performed first. While this makes sense, it also means those with less urgent conditions get continuously pushed back, and with longer waiting times come more adverse patient outcomes. Surgeries that are delayed for too long can often lead to compromised quality of life for patients (Marquedant, 2022). Among the most concerning possibilities is delayed cancer detection, in which the delays could leave larger proportions of cancers that are too advanced to resect or otherwise cure (Wiebe et al., 2022).

By consistently prioritizing urgent surgeries over all others, two waiting lists grow. The first is that of urgent cases which are individually ranked by necessity amongst themselves, and the second is scheduled elective surgeries (Wiebe et al., 2022). Some of these elective surgeries end up becoming urgent as time passes as well. This includes the aforementioned cases of cancer detection and resection (Wiebe et al., 2022). Another area of concern with elective surgery backlogs pertains to pediatric patients who require time-sensitive treatments and surgeries. Some cases are time-sensitive not because of the physical urgency of the surgery, but rather because the pediatric patients may need to keep up with developmental milestones, meaning that delayed surgeries may have long-lived repercussions (Wiebe et al., 2022).

Surgical waiting times did improve at least somewhat since the start of the pandemic. Cancer surgeries, for example, have rebounded in volume between April and September of 2021 to about pre-pandemic levels after falling about 20% in the first 6 months of the pandemic (CIHI, 2022). During this timeframe, other urgent treatments such as radiation therapy and hip fracture surgeries also bounced back to safe levels, with 85% and 97% of patients receiving appropriate treatment in the recommended timeframes, respectively (CIHI, 2022). Several other surgical areas, however, failed to rebound until much further into the pandemic. Knee replacements performed within the recommended time frame, for example, only rose to 59% and hip replacements to 75% between April and September of 2021, leaving many with a decreased quality of life for longer (CIHI, 2022). Some areas of treatment were even able to rebound to levels higher than those seen before the pandemic. MRI scan waiting times were reduced by 5-6 days on average compared to pre-pandemic levels. Such instances of improvement are due mainly to shifts in resources such as increases in staffing which meant MRIs could be performed more frequently and at extended hours of the day (CIHI, 2022).

Much like the case with healthcare worker burnout, unsustainable waiting times for surgeries were not caused by the pandemic: existing issues were merely exacerbated by it. The Fraser Institute found that in 2019 before the COVID-19 pandemic, Canadians had to wait approximately 20.9 weeks for elective treatment between receiving a referral from a practitioner to actually getting treatment (Moir & Barua, 2022). This figure represents an average across the provinces and territories, with significant provincial and territorial differences. Ontario had the shortest figure of 18.5 weeks, and Nova Scotia reported the longest of 53.2 weeks (Moir & Barua, 2022).

Potential Solutions and Next Steps

While the crash of the healthcare system undoubtedly has and will continue to have a detrimental impact on the health and lives of millions of people, the pandemic has also shined a light on the necessity of improvement in many areas of healthcare. For one, the impact on surgical waitlists has raised serious concerns regarding the entire design of Canada's urgency prioritization system. The rapid acceleration of healthcare worker burnout has also revealed weaknesses in the structure of healthcare employment.

Healthcare leaders must address healthcare worker burnout by making innovative adjustments to pre-existing protocols to prioritize the mental health of this group. Firstly, it has been recommended that decision-makers in policy strengthen their relationships with key stakeholders such as clinicians (McEvoy & Thompson, 2022). In this way, they can better understand their experiences and what policy adjustments would make tangible differences in their quality and quantity of work. Secondly, post-traumatic stress disorder is often associated with the arduous and stressful profession of healthcare workers; thus, preventative and awareness programs need to be implemented to address this issue (McEvoy & Thompson, 2022). Thirdly, by focusing on innovation within health-related technology, significant administrative burdens can be lifted off the shoulders of healthcare professionals, allowing them to better manage their responsibilities without experiencing work overload (McEvoy & Thompson, 2022). Fourthly, an emphasis on preventative strategies has also been suggested as a means of promoting mental health well before burnout occurs. Furthermore, it would be beneficial to conduct research on Canadian healthcare environments to single out the factors that contribute to burnout and use that data in the creation of solutions (McEvoy & Thompson, 2022).

In order to properly ensure Canada's readiness for either the next wave of COVID-19 or a new pandemic altogether, its ICU capacity must grow. Ideally, each province should increase the number of staffed ICU beds to increase the overall capacity for ICU admissions (Gibney et al., 2022). This would not only help with accommodating all COVID-related admissions but may assist with the backlog issue as well. If ICU facilities were to be increased, this would mean that appropriate funding would also have to be allocated for the additional staffing needs to cover the capacity (Wiebe et al., 2022). Even before the pandemic, ICU occupancy reached and lingered around 100%. Subsequently, Canada should aim to have enough facilities to operate under and up to 80% occupancy with the ability to handle up to 200% of regular occupancy during times of unregular distress, such as another pandemic (Gibney et al., 2022).

In terms of dealing with the massive backlog of surgeries and treatments, Canada's healthcare system may require a full-scale redesign in areas such as urgency prioritization. There is a necessary balance between improving accessibility to scheduled surgeries and still giving enough attention to urgent cases. The two main notions in a potential restructuring of the prioritization process are that "urgency" is a relative concept, and "scheduled" and "elective" do not mean "optional". Removal of the association between the terms "scheduled" and "optional" would encourage healthcare systems to allocate appropriate capacity for scheduled cases (Wiebe et al., 2022). As well, the relativity of urgency in cases should play a larger role in the performance of scheduled surgeries. For example, urgent cases that can wait until later in their urgency window without consequence should not precede already-scheduled surgeries. This would allow for stronger predictability in scheduling elective surgeries (Wiebe et al., 2022).

Furthermore, more attention should be paid to the consequences of delaying elective surgeries. As mentioned earlier, many elective surgeries have consequences that, while not life-threatening, may lead to serious detriments to quality of life and may complicate the surgery required. Magnifying the significance of these factors would provide a basis for distributing the limited surgical time available over a wider range of cases compared to when just abiding by traditional urgency (Wiebe et al., 2022). Another way that scheduled surgeries could be better supported is by prioritizing scheduling during seasons when the demands for hospital beds and ICU facilities are predictably lower. This would mean scheduling elective surgeries outside of particularly risky times such as flu seasons, which could result in fewer delays and cancellations (Wiebe et al., 2022). Efforts could also be made to increase the efficiency in case booking. Strategies including machine learning algorithms could be used to account for specific variables like individual surgeons and case types (Wiebe et al., 2022). This could effectively help fill the gap in personnel shortages in specific fields by distributing resources in the most efficient ways possible.

Regardless of the type of change, the constant between the options is the need for further government funding. In order to clear the bottleneck of surgeries still in a backlog and to prevent similar backlogs from happening in the future and potentially crashing the healthcare system, governments must increase the amount of resources allocated to healthcare providers and urge them to challenge the status quo and create meaningful change to the healthcare system (Wiebe et al., 2022). Ideally, if appropriate changes are made, they could clear up built-up issues such as the backlog of surgeries and severe burnout in clinicians and could better prepare Canada for the inevitable next pandemic.

Conclusion

All in all, the COVID-19 pandemic undoubtedly left devastating damage on Canada's healthcare systems and those of most countries around the world. Through the physical and mental overworking of healthcare workers, the overflow of ICU admissions and the overwhelming backlogs of both urgent and elective surgeries, many existing weaknesses in Canadian healthcare have been brought to the surface. It is important to recognize that many of these issues existed long before the pandemic, and due to a worldwide crisis, they were exacerbated beyond the point of ignoring.

References

Appleby, J. (2022). The public finance cost of covid-19. BMJ, 376, o490. https://doi.org/10.1136/bmj.o490

CIHI. (2021, November 4). COVID-19 expected to push Canada's health spending beyond $300 billion in 2021 I CIHI. https://www.cihi.ca/en/news/covid-19-expected-to-push-canadas-health-spending-beyond-300-billion-in-2021

CIHI. (2022, May 10). Patients in Canada continued to experience longer wait times for surgeryduring COVID-19 pandemic I CIHI.https://www.cihi.ca/en/news/patients-in-canada-continued-to-experience-longer-wait-times-forsurgery-during-covid-19

Gibney, R. T. N., Blackman, C., Gauthier, M., Fan, E., Fowler, R., Johnston, C., Jeremy Katulka, R., Marcushamer, S., Menon, K., Miller, T., Paunovic, B., & Tanguay, T. (2022). COVID-19 pandemic: The impact on Canada's intensive care units. FACETS, 7, 1411–1472 https://doi.org/10.1139/facets-2022-0023

ICN. (2022, May 12). "The greatest threat to global health is the workforce shortage"—International Council of Nurses International Nurses Day demands action on investment in nursing, protection and safety of nurses. ICN - International Council of Nurses. https://www.icn.ch/news/greatest-threat-global-health-workforce-shortage-international-council-nurses-international

Marquedant, K. (2022, August 18). Surgical Backlogs From COVID-19 Persist and Could Have Serious Healthcare Consequences Going Forward. Massachusetts General Hospital. https://www.massgeneral.org/news/press-release/surgical-backlogs-covid-19-persist-serious-consequences

McEvoy, J., & Thompson, S. (2022, June 27). Healthcare burnout in Canada: Facing the problem and creating solutions to better support our healthcare workers I NATIONAL. NATIONAL. https://www.national.ca/en/perspectives/detail/healthcare-burnout/

Moir, M., & Barua, B. (2022, April 27). Long wait times for health care predated pandemic: Op-ed. Fraser Institute. https://www.fraserinstitute.org/article/long-wait-times-for-health-care-predated-pandemic

Murthy, V. H. (2022). Confronting Health Worker Burnout and Well-Being. New England Journal of Medicine, 387(7), 577–579. https://doi.org/10.1056/NEJMp2207252

Richards, F., Kodjamanova, P., Chen, X., Li, N., Atanasov, P., Bennetts, L., Patterson, B. J., Yektashenas, B., Mesa-Frias, M., Tronczynski, K., Buyukkaramikli, N., & El Khoury, A. C.(2022). Economic Burden of COVID-19: A Systematic Review. ClinicoEconomics and Outcomes Research: CEOR, 14, 293–307. https://doi.org/10.2147/CEOR.S338225

Robinson, L. (2022, October 18). 2022 State of Health-care Performance Improvement Report | Kaufman Hall. KaufmanHall. https://www.kaufmanhall.com/insights/research-report/2022-state-healthcare-performance-improvement-report

Wiebe, K., Kelley, S., & Kirsch, R. E. (2022). Revisiting the concept of urgency in surgical prioritization and addressing backlogs in elective surgery provision. CMAJ, 194(29), E1037–E1039. https://doi.org/10.1503/cmaj.220420

Public Health Recommendations: What worked?

By Kanish Baskaran

"Flattening the curve" … "Reducing transmission" … "Protecting the vulnerable". Phrases like these were all too common over the last few months as Governments around world moved at warp-pace to contend with the COVID-19 pandemic. While the pandemic has affected different countries to different degrees, almost universally, the disease has put a strain on healthcare systems and the economy as a whole. Governments have worked to implement measures that slow the spread of the virus, and protect their populations, while simultaneously curtailing any sociocultural and economic effects. These strategies have included various measures, including lockdowns, mask mandates and vaccination campaigns.

Another key impact of the COVID-19 pandemic was that it brought sociocultural, structural issues and problems to light, that had been building in Western nations like Canada for many years prior. This includes the decrease in public trust in both government and experts, as highlighted by the continuous wave of insults aimed at public health officials. These issues have created new twists in Healthcare problems and policy, that may take years to reach a consensus opinion.

2 main features were stressed by governments in their pandemic response: Speed and Scale (Ayouni et al., 2021). As

many may have heard on the news and other such sites, epidemiologists often seek to "flatten the [epidemic] curve". This effectively refers to delaying, as well as decreasing the intensity of the pandemic peak, providing adequate time for other interventions to be developed that effectively treat the condition (Ayouni et al., 2021) (e.g. vaccines, treatments).

Public Health Strategies used during the Pandemic

Lockdowns

One of the primary strategies employed to slow the spread of the virus has been the implementation of lockdowns, which involve the closure of non-essential businesses and the restriction of movement and social gatherings (Chen et al., 2021). These measures are designed to reduce the number of contacts between people, thereby reducing the transmission of the virus (Chen et al., 2021). While lockdowns can be effective in slowing the spread of the virus, they can also have negative impacts on the economy and mental health of the population. Therefore, it is important for governments to carefully consider the balance between the benefits and drawbacks of lockdowns and to provide support to those who are adversely affected by them (Chen et al., 2021).

Mask Mandates

Another key strategy in the fight against COVID-19 has been the use of face masks (Horney, 2022). Wearing a mask can help to reduce the spread of the virus by trapping droplets that are produced when an infected person talks, coughs, or sneezes (Horney, 2022). Masks can be particularly effective in reducing the transmission of the virus in crowded or enclosed

44

spaces where it is difficult to maintain social distance (Horney,

2022). Some countries have implemented mask mandates, requiring people to wear masks in certain settings or situations, while others have encouraged the use of masks through public health campaigns and education.

Vaccination campaigns

Vaccination campaigns have played a crucial role in the response to the COVID-19 pandemic, with several vaccines now available and in use around the world (Lina, 2020). Vaccines work by exposing the body to a small, harmless part of a virus or bacteria, which triggers the immune system to produce antibodies that can fight off future infections (Wilder-Smith, 2021). COVID-19 vaccines have been developed and deployed at an unprecedented speed, with several vaccines now available and in use around the world (Polisena et al., 2021).

Vaccination campaigns have been implemented by governments and health agencies in order to protect the population from the virus and reduce the number of cases and deaths from COVID-19 (Christen, 2022). These campaigns have typically targeted vulnerable populations such as the elderly and those with underlying health conditions, as well as essential workers and other high-risk groups (Christen, 2022). In many cases, vaccination has been made available to the general population as well.

Vaccination campaigns have typically involved the distribution of vaccines to vaccination sites, such as clinics, hospitals, and pharmacies (Tabari et al., 2020). These sites may be run by public health agencies or by private organizations and may offer vaccines on a walk-in basis or by appointment

(Wilder-Smith, 2021). In some cases, mobile vaccination units have been used to bring vaccines to remote or underserved areas.

In order to ensure that vaccines are distributed efficiently and effectively, vaccination campaigns have typically involved the use of advanced logistics and supply chain management systems (Ayouni et al., 2021). These systems have been used to track the movement of vaccines from manufacturers to distribution centers, and then to vaccination sites (Ayouni et al., 2021). They have also been used to manage the distribution of supplies such as syringes and personal protective equipment (PPE).

Countries need to prioritize the distribution of limited vaccine supplies. This has required governments and health agencies to make difficult decisions about which groups should receive the vaccine first (Dyer, 2022). In many cases, priority has been given to vulnerable populations and essential workers, with the general population being vaccinated at a later stage (Dyer, 2022).

Another challenge has been the need to ensure that vaccines are administered safely and effectively. This has required the training and deployment of large numbers of healthcare workers and other personnel to administer the vaccines (Talic et al., 2021). It has also involved the development of robust systems for monitoring the safety and effectiveness of the vaccines, including systems for tracking, and reporting adverse events (Talic et al.,2021).

Contact Tracing

Contact tracing is a public health strategy that involves identifying and tracking the contacts of infected individuals in order to alert them to the risk of infection and to provide them with guidance on how to protect themselves (Dyer, 2022). It is an important tool in the fight against the COVID-19 pandemic, as it helps to reduce the spread of the virus by identifying and isolating individuals who may have been exposed to the virus (Dyer, 2022).

Contact tracing involves identifying the close contacts of infected individuals, which are defined as those who have spent 15 minutes or more within 6 feet of an infected person (Horney, 2022). Close contacts are considered to be at high risk of infection and are advised to self-quarantine and monitor for symptoms (Horney, 2022). Contact tracers will also gather information about the movements and activities of infected individuals in order to identify other potential contacts who may have been exposed to the virus (Horney, 2022).

Contact tracing is typically carried out by trained public health professionals, who use a variety of methods to identify and track contacts (Polisena et al., 2021). These methods may include phone calls, emails, or text messages, and may involve the use of specialized software or apps (Talic et al., 2021). In some cases, contact tracers may also use more traditional methods such as door-to-door outreach or in-person interviews.

One of the main challenges of contact tracing is the need to identify and track a large number of contacts in a short period of time (Dyer, 2022). This can be particularly difficult in the early stages of an outbreak, when the number of cases may be rapidly increasing and the capacity of contact tracing systems may be stretched. In order to be effective,

47

contact tracing systems need to be well-coordinated and have sufficient resources, including trained staff and technology.

Another challenge of contact tracing is the need to ensure the privacy and confidentiality of infected individuals and their contacts (Dyer, 2022). This can be particularly sensitive in cases where the infected individual may be reluctant to share information about their contacts or activities. In order to address these concerns, contact tracing systems shouldbe designed with robust privacy and confidentiality protections in place.

Country differences in addressing spread of COVID–19

China

One country that has been at the centre of controversy for their public health policies resulting from the COVID-19 pandemic has been China (Dyer, 2022). From the start of the Pandemic, this nation has taken drastic measures to prevent the outbreak from spreading – called the Zero COVID policy (Chen et al., 2021). This included quarantining entire cities, imposing strict travel bans, and even detaining individuals that break lockdown rules (Chen et al., 2021). This has often been described as a double-edged sword however, because while this can prevent the spread of the pandemic, when not paired with an effective vaccination strategy, this can cause economic recessions and mass public dissent (Chen et al., 2021). The Zero COVID approach aimed to prevent viral transmission through a variety of different measures, which includes (but is not limited to) contact-tracing, quarantines, and vaccination. This approach occurs in 2 different steps usually, with the first being the initial containment phase to remove the virus from the region, and the second being the sustained containment phase, where the virus is prevented from re-entering the community (Chen et al., 2021). Elimination strategies are contrasted to mitigation strategies, whose goal is to lessen the impacts of the disease, but still tolerating some level of transmission.

The double-edged nature of a Zero COVID policy is exemplified in the case of China, which was initially celebrated for its ability to overcome the pandemic in a timely fashion through this policy. However, in recent months, the Communist party has faced increased dissent both internally and abroad regarding its zero-policy strategy, which had held its economy back while also worsening the mental state of its citizens (Chen et al., 2021). This was worsened by reports of a fire burning down a locked building in Xinjiang, which could have been avoided if officials did not have to obey social distancing rules.

Strategies to fight disinformation used during the pandemic

Disinformation, or false information that is spread deliberately to deceive, has been a significant problem during the COVID-19 pandemic (Talic et al., 2021). It can take many forms, including conspiracy theories, misinformation about the virus and its origins, and false cures or prevention methods (Talic et al., 2021). Disinformation can have serious consequences, as it can undermine public trust in authorities and experts, and lead to harmful behaviors such as refusal to wear masks or get vaccinated (Talic et al., 2021). Some strategies used to fight disinformation used by governments are listed below;

Promoting accurate and reliable information

One of the most important strategies to fight disinformation is to provide accurate and reliable information to the public. This can be done through official sources such as government websites and health agencies, as well as trusted media outlets and fact-checking organizations (Lina, 2020). It is important

to make sure that the information is easy to understand and accessible to all members of the community (Lina, 2020).

Educating the public

Disinformation is a major problem that can have serious consequences for individuals and society as a whole. It can lead to the spread of false information and harmful beliefs, and can even be used to manipulate public opinion and interfere in elections (Christen, 2022). Therefore, it is important that governments take steps to educate the public about how to identify and combat disinformation (Christen, 2022).

One way that governments can do this is by providing resources and information to help people learn how to spot disinformation. This can include tips on how to evaluate the credibility of sources, how to fact-check information, and how to recognize common tactics used to spread disinformation (Horney, 2022). Governments can also work with schools and other educational institutions to incorporate lessons on media literacy and critical thinking into the curriculum, so that people can learn these skills from an early age (Horney, 2022).

In addition to providing education, governments can also take a more proactive approach by implementing policies and regulations that aim to combat disinformation (Polisena et al., 2021). This can include measures such as requiring social media platforms to label or remove false or misleading content, or imposing fines on organizations that engage in disinformation campaigns (Polisena et al., 2021).

Ultimately, the key to effectively combating disinformation is to create a society that is informed and able to think critically about the information it consumes (Dyer, 2022). By educating the public and implementing policies that discourage the spread of disinformation, governments can help to create a more informed and engaged citizenry.

Utilizing social media Platforms to address disinformation

Social media platforms have a significant role in the spread of disinformation, and it is important to work with these platforms to address the issue (Unruh et al., 2022). This can involve reporting disinformation and working with the platform to remove it, as well as using algorithms and other tools to reduce the spread of misinformation (Unruh et al., 2022). For example, many governments have set up official accounts on platforms like Twitter and Facebook and use these accounts to share updates and information about the pandemic. They may also use these accounts to respond to questions from the public or to provide links to reliable sources of information (Unruh et al., 2022). In addition to using social media to share information, some governments have also implemented policies or regulations to address disinformation on these platforms. This can include measures such as requiring social media companies to remove false or misleading content, or imposing fines on individuals or organizations that engage in disinformation campaigns (Polisena et al., 2021).

Encouraging responsible reporting

Media organizations have a responsibility to report accurate and reliable information. Encouraging responsible reporting can involve providing training and resources for journalists, as well as holding media organizations accountable for spreading disinformation (Horney, 2022).

Working with community leaders

Community leaders, such as religious leaders, teachers, and local politicians, can play a crucial role in fighting disinformation.

By working with these leaders, it is possible to reach a wider audience and to provide accurate information in a way that is more trusted and believable (Ayouni et al., 2021).

Encouraging dialogue and debate

Encouraging dialogue and debate can be an effective way to challenge disinformation. By providing a forum for people to discuss and question information, it is possible to encourage critical thinking and to expose disinformation for what it is (Unruh et al., 2022).

Strategies for nations to work together

Shared Vaccination programs

One effective measure to evolve out of the COVID-19 pandemic has been shared vaccination programs (Horney, 2022). These programs are designed to ensure that vaccines are distributed fairly and equitably around the world, so that everyone has the opportunity to be protected against the virus (Talic et al., 2021).

One of the main shared vaccination programs that has been implemented during the pandemic is the COVAX Facility (Talic et al., 2021). This initiative is led by the World Health Organization (WHO) and aims to provide vaccines to low- and middle-income countries that may not have the resources to purchase them directly (Horney, 2022). The COVAX Facility is funded through donations from governments and other organizations and has worked to secure agreements with vaccine manufacturers to purchase and distribute doses to participating countries (Unruh et al., 2022).

In addition to the COVAX Facility, many countries have also established their own shared vaccination programs to ensure that their populations have access to vaccines. For example, some countries have set up vaccination programs for their own citizens and have also donated or provided vaccines to other countries in need (Chen et al., 2021).

World Health Organization

The World Health Organization (WHO) has played a crucial role in coordinating the global response to the COVID-19 pandemic. As the United Nations' specialized agency for health, the WHO has a mandate to provide leadership on global health matters and to work with member states to ensure the highest possible level of health for all people (Chen et al., 2021).

In the early days of the pandemic, the WHO worked to provide guidance to countries on measures to contain the spread of the virus and treat those who were infected. This included providing recommendations on measures such as social distancing, the use of masks, and the isolation of cases. The WHO also worked to ensure that countries had access to the necessary supplies and equipment to deal with the pandemic, such as personal protective equipment (PPE) and ventilators (Unruh et al., 2022).

The WHO has also played a key role in the development and distribution of COVID-19 vaccines. It has worked with vaccine manufacturers to ensure that vaccines are safe, effective, and of high quality, and has provided guidance on the distribution of vaccines (Shaikh et al., 2021). As mentioned previously, the WHO has also led the COVAX facility, an international initiative to ensure equitable access to COVID-19 vaccines, which has secured more than two billion doses of vaccines and provided them to more than 170 countries (Dyer, 2022).

Sharing of Information and Research

The sharing of research has played a vital role in the global response to the COVID-19pandemic (Polisena et al., 2021). Scientists and medical professionals from around the world have come together to share their findings and knowledge about the virus, in an effort to better understand it and develop effective treatments and vaccines (Unruh et al., 2022). One way that research has been shared during the pandemic is through the publication of scientific papers and articles in peer-reviewed journals. These articles provide detailed descriptions of research findings and are subject to review by other experts in the field before being published (Ayouni et al., 2021). This ensures that the research is of high quality and can be relied upon by other scientists and medical professionals (Musolff (editor) et al., 2022).

In addition to traditional methods of sharing research, the COVID-19 pandemic has also seen the use of new technologies and platforms to facilitate the sharing of research. For example, many researchers have made their data and findings available through online platforms such as preprint servers, which allow researchers to share their work before it has been peer-reviewed (Horney, 2022). This has allowed for the rapid dissemination of research findings and has helped to speed up the development of treatments and vaccines.

The sharing of research has been crucial in the development of COVID-19 vaccines. Scientists from around the world have worked together to identify the genetic sequence of the virus and to develop vaccines that are effective against it (Horney, 2022). This cooperation has been vital in the development of multiple vaccines that have been distributed globally.

Pandemic Preparedness Strategies going forward

Going forward, it is important for countries to be better pre-pared for global pandemics in order to minimize their impact and better protect the health and well-being of people around the world. There are a number of measures that countries can take to improve their pandemic preparedness:

1. Strengthening health systems: This includes investing in the infrastructure and resources needed to quickly identify and respond to outbreaks, such as laboratories, health work-er training, and emergency preparedness plans (Chen et al., 2021).

2. Enhancing surveillance and response: This involves implementing systems to monitor and track outbreaks, as well as developing plans for how to respond to them. This could include measures such as quarantine, testing, and contact tracing (Polisena et al., 2021).

3.   Developing and stockpiling vaccines and other medical supplies: Having a supply of vaccines and other medical supplies on hand can help to quickly contain outbreaks and protect the population (Ayouni et al., 2021). It is important to invest in the research and development of new vaccines and to have adequate stockpiles of existing vaccines (Horney, 2022).

4.   Improving international coordination: In the event of a global pandemic, it is important for countries to work together and share information and resources. This could include further establishing international agreements to exchange

medical supplies and expertise, or to aid each other in times of need (Unruh et al., 2022).

5.   Promoting research and development: Investment in research and development is crucial in order to better understand emerging diseases and to develop effective treatments and vaccines (Chen et al., 2021). This could include funding research on the development of new technologies or therapies, as well as investing in the infrastructure needed to support this research (Unruh et al., 2022).

References

Ayouni, I., Maatoug, J., Dhouib, W., Zammit, N., Fredj, S. B., Ghammam, R., & Ghannem, H. (2021). Effective public health measures to mitigate the spread of COVID-19: A systematic review. BMC Public Health, 21, 1015. https://doi.org/10.1186/s12889-021-11111-1

Chen, H., Shi, L., Zhang, Y., Wang, X., Jiao, J., Yang, M., & Sun, G. (2021). Response to the COVID-19 Pandemic: Comparison of Strategies in Six Countries. Frontiers in Public Health, 9. https://www.frontiersin.org/articles/10.3389/fpubh.2021.708496

Christen, P. (2022). How do countries' responses to the health effects of Covid-19 compare? Economics Observatory. https://www.economicsobservatory.com/how-do-countries-responses-to-the-health-effects -of-covid-19-compare

Dyer, O. (2022). Covid-19: Canada outperformed comparable nations in pandemic response, study reports. BMJ, 377, o1615. https://doi.org/10.1136/bmj.o1615

Horney, J. (2022). The COVID-19 Response: The Vital Role of the Public Health Professional. Academic Press. http://gen.lib.rus.ec/book/index.php?md5=5EE71F13EFC494B-18F239E63C9F2FDB0

Lina. (2020, March 20). Comparing international responses to COVID-19. Global Health. https://globalhealth.stanford.edu/covid-19/comparing-international-responses-to-covid-19.html/

Musolff (editor), A., Breeze (editor), R., Kondo (editor), K., & Vilar-Lluch (editor), S. (2022). Pandemic and Crisis Discourse: Communicating COVID-19 and Public Health Strategy. Bloomsbury Academic. http://gen.lib.rus.ec/book/index.php?md5=1838373E0E9D8CE37E1A80F-F7656A7D2

Polisena, J., Ospina, M., Sanni, O., Matenchuk, B., Livergant, R., Amjad, S., Zoric, I., Haddad, N., Morrison, A., Wilson, K., Bogoch, I., & Welch, V. A. (2021). Public health measures to reduce the risk of SARS-CoV-2 transmission in Canada during the early days of the COVID-19 pandemic: A scoping review. BMJ Open, 11(3), e046177. https://doi.org/10.1136/bmjopen-2020-046177

Shaikh, N. F., Kunjir, A., Shaikh, J., & Mahalle, P. N. (2021). COVID-19 Public Health Measures: An Augmented Reality Perspective (1st ed.). CRC Press. http://gen.lib.rus.ec/book/index.php?md5=E0A-B806159A9D20139D3230DB3B28453

Tabari, P., Amini, M., Moghadami, M., & Moosavi, M. (2020). International Public Health Responses to COVID-19 Outbreak: A Rapid Review. Iranian Journal of Medical Sciences, 45(3), 157–169. https://doi.org/10.30476/ijms.2020.85810.1537

Talic, S., Shah, S., Wild, H., Gasevic, D., Maharaj, A., Ademi, Z., Li, X., Xu, W., Mesa-Eguiagaray, I., Rostron, J., Theodoratou, E., Zhang, X., Motee, A., Liew, D., & Ilic, D. (2021). Effectiveness of public health measures in reducing the incidence of covid-19, SARS-CoV-2 transmission, and covid-19 mortality: Systematic review and meta-analysis. BMJ, 375, e068302. https://doi.org/10.1136/bmj-2021-068302

Unruh, L., Allin, S., Marchildon, G., Burke, S., Barry, S., Siersbaek, R., Thomas, S., Rajan, S., Koval, A., Alexander, M., Merkur, S., Webb, E., & Williams, G. A. (2022). A comparison of 2020 health policy responses to the COVID-19 pandemic in Canada,

Ireland, the United Kingdom and the United States of America. Health Policy, 126(5), 427–437. https://doi.org/10.1016/j.healthpol.2021.06.012

Wilder-Smith, A. (2021). COVID-19 in comparison with other emerging viral diseases: Risk of geographic spread via travel. Tropical Diseases, Travel Medicine and Vaccines, 7(1), 3. https://doi.org/10.1186/s40794-020-00129-9

Vaccine Development and Clinical Trials

By Tristan Ramsubag

Vaccines are an effective form of active acquired immunity against viruses. An individual receives a viral/immune stimulus or antigen (ex. viral proteins, genetic materials, etc.) that activates lymphocytes or specialized immune cells known as B cells and T cells. These cells process the antigen to produce proteins called antibodies. The body's immune system uses the antibodies to quickly recognize, target, attack, and eliminate the virus efficiently. The process prevents the progression of the infection. Vaccines introduce the immune stimulus directly into the patient's body; however, the antigens are processed via diverse technologies to ensure it does not cause an infection in the host.

The SARS-CoV-2 pandemic was immensely detrimental to public health systems worldwide. It inflicted adverse global consequences and caused millions of infections and deaths. Its impact is still ongoing. Traditionally, vaccine development and clinical trials are lengthy and tedious. However, the vaccine production and clinical trials were unprecedently accelerated to respond effectively to the Covid-19 pandemic. The development of Covid-19 vaccines involve building on existing knowledge and using new innovative technologies. Covid-19 vaccines deviated from the traditional approaches, thus, shifting the vaccinology paradigm and sparking widespread clinical interest and research (Ndwandwe & Wiysonge, 2021).

Covid–19 Vaccine Development

There are four categories of Covid-19 vaccines. Each utilizes diverse viral components to act as antigens and employs new methods of production. The categories encompass whole virus vaccines, protein-based vaccines, viral vector vaccines, and nucleic acid vaccines.

Whole Virus Vaccines

Whole virus vaccines stimulate the individual's protective immunity by introducing a weakened or attenuated form of the Covid-19 virus. Since this type of vaccine uses viruses that are inactivated or dead, it does not cause the disease. Live attenuated and inactivated vaccines are two types of whole virus vaccines (Ndwandwe & Wiysonge, 2021). Live attenuated SARS-CoV-2 vaccines replicate and grow in the body like the original form of the virus. To create these vaccines, scientists culture the virus repeatedly by selectively using strains that are less virulent within the previous cultures. This attenuates the virus so that it is not capable of causing the disease but elicits an immune response. Inactivated vaccine production involves subjecting vial cultures to heat, chemicals and/or radiation to destroy genetic material and kill the viruses. The virus within inactivated vaccines cannot replicate (Denizer et al., 2018); (Mijanur et al., 2022). This form of vaccine development is conventional. It is the basis for many existing vaccines that are commonly administered, including measles, polio, and influenza (Ndwandwe & Wiysonge, 2021). Both types work through different mechanisms (ex. humoral vs cellular immunity). Eight Covid -19 vaccines use this technology with approval in over 60 countries. Sinopharm, BBIBP-CorV, and CoronaVac are a few examples of whole virus Covid-19 vaccines (Mijanur et al., 2022).

Protein Subunit Vaccines

The vaccine contains specific fragments found in the Covid-19 virus. The fragments act as antigens and trigger an immune response. Proteins, polysaccharides, or other cellular components isolated from the virus are used to develop the vaccine. Many Covid-19 vaccines are protein-based. Proteins can be isolated from cultures. It can also be produced using recombinant genetic engineering that clones DNA and reproduces the protein in large quantities. Other Covid-19 vaccines use virus-like particles (VLP). This form of vaccine incorporates empty viral protein shells. It is based on the finding that the expression of certain proteins spontaneously arranges into structures found in viral pathogens. The shells mimic the structure of the virus and stimulate the host's immune system. It is not infectious because it is absent of genetic material. The EpiVacCorona, Abdala, and MVC-COV vaccines are examples of protein-based vaccines. Currently, there are numerous other protein subunit Covid-19 vaccines in the clinical trial stages (Denizer et al., 2018); (Mijanur et al., 2022).

Viral Vector Vaccines

Viruses cause an infection by attacking and injecting viral genetic material into human cells. Viruses use the human genetic and protein synthesis processes to replicate, thus, causing an infection. Viral vector vaccines work based on this concept. In viral vector vaccines, a modified virus is used to deliver a genetic code into the host. The genetic material is injected into the host's cells to make antigens and initiate an immune reaction. The modification of the vector ensures that it does not result in a disease as it is weakened. Covid vaccine using this technology generates the SARS-CoV-2 spike proteins within the host. Covishield from AstraZeneca is produced from this method of vaccinology (Mijanur et al., 2022).

Nucleic Acid Vaccines

The first nucleic acid vaccines were manufactured to provide immunity against the Covid-19 virus. The two types are DNA-based and RNA-based vaccines. The DNA or RNA segments that code for a specific viral antigen are inserted into a bacterial plasmid and then into the human host. The host cellular machinery expresses the viral genetic material and produces the antigen. The antigen will trigger an immune response. The Moderna and Pfizer Covid-19 vaccines are mRNA-based vaccines, and the first nucleic acid drugs or vaccines licensed for human administration (Mijanur et al., 2022).

Covid–19 Clinical Trials

Vaccines are subjected to a series of clinical trials and phases before being approved by federal regulatory bodies for public use. The pre-clinical stage involves conducting basic scientific laboratory and animal studies. This stage usually takes 5 – 10 years. Next is the clinical stage with 3 phases. The vaccine is tested on a small sample of healthy individuals in phase one. The goal of phase 1 is to determine dosage and pharmacokinetics. Conventionally, this process takes about 3 years. Phases 2 and 3 involve testing the vaccine on a larger and more representative sample to assess safety, efficacy, and side effects. The process takes about 3 years each. The monitoring stage or phase 4 requires post-market surveillance. Clinical trials for Covid-19 vaccines were accelerated. It took 6 - 9 months for vaccines to progress from the pre-clinical stage to the end of phase 1. Phases 2 to 3 also lasted 6 – 9 months (Ndwandwe & Wiysonge, 2021). WHO and FDA's minimum criteria for acceptability is the demonstration of a minimum of 50% vaccine efficacy. Vaccine efficacy or effectiveness is a measure of a vaccine's ability to prevent an infection. For instance, a 50% vaccine efficacy means that the vaccine will prevent 50% of people from contracting the disease when in contact with a pathogen. WHO standards are based on disease progression and transmission, while FDA examines laboratory confirmed cases (Kwok, 2021). In the early stages of the Covid-19 vaccine development, studies using nonhuman primates identified functional antibodies protecting against the coronavirus (Barouch, 2022).

Pormohammad et al. (2021) presented a meta-analysis comparing the safety and efficacy of Covid-19 vaccines based on the data reported in randomized clinical trials. Nucleic acid vaccines, specifically mRNA vaccines, reported the highest efficacy. These vaccines produced an efficacy above 94% after the second dose. Adenovirus-vectored vaccines produced similar results. Whole viruses and protein-based vaccines produced about 70% to 90% efficacy following the phase 3 trials.

The BioNTech-Pfizer-mRNA and the Moderna mRNA-1273 vaccines were the principal vaccines administered to tackle the Covid-19 pandemic in Canada. During phase 3 of the clinical development, the Pfizer vaccine was tested on 43548 subjects. The group receiving the mRNA vaccine from Pfizer and BioNTech consisted of 17411 participants and the control group had 17511 participants. There were 8 confirmed Covid-19 cases in the group receiving the vaccine, and 162 cases in the control group, producing a 95% vaccine efficacy. Reactions to the vaccine were mild to moderate. No severe or life-threatening adverse reactions were reported in the clinical study. Reported side effects included fever and chills, however, it only lasted 1 to 2 days (Kwok, 2021). Moderna produced vaccines containing the mRNA-1273. In phase 1, animal studies were used to test mRNA-1273. The RNA segment codes for the S-2P antigen which is a component of a viral trans-membrane protein. Phase 1 testing revealed antiviral properties of the mRNA-1273 in all test subjects with no evidence of safety concerns. A phase 3 study had 30420 participants. There were only 11 cases in the experimental group receiving the vaccine, and 185 cases of Covid-19 in the control group, producing a vaccine efficacy of 94% (Kwok, 2021).

The pharmacokinetics of the Covid-19 vaccines may differ among populations, resulting in deviations from the vaccines' safety and efficacy reported in the clinical trials. For instance, many phase 3 clinical studies were limited regarding representation. Populations comprising pregnant women, critically ill patients, the elderly, etc. were excluded from phase 3 studies. More studies focusing on specific groups of individuals are emerging to address this limitation.

Vaccines Against Variants

Viruses are consistently evolving, evident by the patterns seen in the coronavirus pandemic. Following the original coronavirus that sparked the pandemic in 2020, the SARS-CoV-2 virus underwent multiple mutations responsible for subsequent waves of infection. The mutations (called D614G substitutions) occurred in the genes coding for spike proteins and conferred a fitness advantage. The variants increased Covid-19 transmissibility. The mutations also helped the virus escape detection from antibodies in some cases, therefore, causing people with acquired immunity to be re-infected (Barouch, 2022). There were four variants of concern during the Covid-19 pandemic: Alpha (B.1.1.7), Beta (B.1.351), Gamma (P.1), and Delta (B.1.617.2). In addition, the Omicron (B.1.1.529) variant is a currently circulating variant of concern. Post-vaccination population studies were conducted to determine the efficacy of vaccines in reducing the rate of symptomatic infections, hospitalizations, and death.

The Alpha variant was first identified in the United Kingdom. It spreads faster than other versions of the virus. Alpha causes more severe symptoms and increases the risk of fatality (Hayawi et al., 2021). Compared to the original virus, Alpha mutations increase viral transmissibility by more than 56 percent. It is associated with a 61 – 64 percent mortality rate (Fiolet, 2022). Fourteen days after the first dose, mRNA-1273 vaccines prevented symptomatic infections in 80 – 84 percent of recipients. There was an 88 – 99 percent efficacy 7 days after the second dose. Moderna's effectiveness against the Alpha strain was slightly higher than Pfizer vaccines. Studies reported an efficacy of 65 – 68 percent (14 days after dose 1) and 87 – 90 percent (7 days after dose 2); (Nasreen, 2022). Overall, studies demonstrate an 88 – 100% effectiveness of all mRNA vaccines in the prevention of the Alpha disease. Full immunizations (2 doses of mRNA vaccines) showed an 89 – 95 percent efficacy at preventing hospitalizations and death after an individual becomes infected with the Alpha variant. Vaccine efficacy against Alpha is reported to be higher than other variants (Fiolet, 2022).

The Beta variant was first identified in South Africa, while the Gamma variant was first identified in Japan and Brazil. Both versions of the Covid-19 virus spread faster than the original pathogen. However, there is no current data indicating that these variants cause an increased severity of symptoms or mortality rate (Hayawi et al., 2021). Transmission rates elevated 50 – 160 percent during waves caused by these variants (but rates were region dependent); (Fiolet, 2022). The variants had a minimal to moderate impact on post - vaccinated populations depending on the vaccine received. First dose protection was 75% for mRNA -1273, which was significantly higher than the 64% protection generated by the Pfizer vaccine. After the second dose, efficacy improved to 96% and 75% respectively (Fiolet, 2022); (Nasreen, 2022). A literature review showed an efficacy of 76 – 100% for mRNA vaccines in terms of preventing symptomatic Beta and Gamma infections. A 95 % protection against Beta and Gamma hospital-izations and deaths was also associated with mRNA vaccines (Fiolet, 2022).

The Delta variant was first identified in India. It spreads faster and causes more severe cases of the disease when compared to the original coronavirus and the other aforementioned variants (Hayawi et al., 2021). Delta transmission was 40% to 60 % higher than the initial form of the coronavirus. It increased hospitalization and death rates worldwide. The Delta variant also caused a major reduction in the efficacy of all Covid-19 vaccines. First dose efficacy was 70% for Moderna and 57% for Pfizer. Second doses elevated efficacy to 95% for Moderna and 92% for Pfizer (Fiolet, 2022); (Nasreen, 2022). As with all variants, mRNA vaccines had the greatest protection against symptomatic Delta infections (47 – 88%). These vaccines generated an 80 – 90% protection against Delta hospitalizations and deaths (Fiolet, 2022).

The Omicron variant was first discovered in South Africa. It is currently the predominantly circulating variant of concern. Compared to all other forms of the Covid-19 virus, the Omicron strain displays high transmissibility and resistance to vaccine and naturally acquired immunity. There are more than 50 mutations, making this variant highly concerning in terms of symptoms, hospitalizations, and deaths. It is responsible for the surge in severe Covid -19 cases in some individuals following the preliminary administration of vaccines. Omicron greatly decreases the efficacy of the primary vaccine doses; though, booster shots are a vital method in preventing this infection (Chenchula, 2022). A study showed that average vaccine efficacy was only 36%, after 7 days of full immunizations or 2 doses of an mRNA vaccine. After 180 days, there was no protection. Protection against Omicron substantially increased to 61% after a third dose or booster shot administration. Therefore, 2 doses provide modest protection against symptomatic Omicron infections while booster doses improve protection. Boosters are especially significant in reducing the risk of severe outcomes following an Omicron infection (Buchan, 2022).

Bivalent mRNA vaccines are currently in phases 2 – 3 of clinical trials. These vaccines contain components of the original virus (which acts as a booster) and the Omicron variant. Therefore, bivalent Covid-19 vaccines result in potent, persistent and broad immune protection/responses to the Covid-19 virus and its multiple mutated strains (including Omicron); (Chalkias et al., 2022).

Conclusion

Vaccines are subjected to rigorous testing to ensure safety and efficacy. The same is also true for Covid-19 vaccines. Clinical trials reveal that Covid-19 vaccines are highly efficient at generating short-term immunity against the virus. The mRNA vaccines display the highest efficacy. Booster shots prevent the waning of and prolong the vaccines' protection. Boosters also enhance the vaccines' efficacy against the coronavirus and its various variants. Studies illustrate that these vaccines reduce the risk of severe symptomatic Covid-19 infections, hospitalizations, and death. Lauring et al., (2022) conducted a study investigating Covid-19 hospitalizations and vaccinations. According to the findings, two doses of mRNA vaccines are extremely effective at reducing the risk of Covid-19 associated hospitalizations caused by the original form of the virus and variants of concern (Alpha and Delta variants). The researchers also demonstrated that three doses of mRNA vaccines were required to produce the same effects against the Omicron variant. To add, the study showed vaccinated patients hospitalized with Covid-19 experienced a lower disease severity than unvaccinated patients. Overall, Covid-19 vaccines are essential and proven tools to protect individuals from the coronavirus infection. However, to build full immunity, individuals must receive the appropriate number of doses. Boosters are also required to maintain sufficient antibody generation and action, and offer adequate protection against variants. There are also special boosters emerging to enhance the protection against variants of concern (ex.Bivalent vaccines).

Vaccines prevent most people from getting sick. Some people may still get infected with the Covid-19 virus but, vaccines will reduce the severity of the disease and prevent its progression into a critical illness requiring hospitalizations. Clinical trail phases 1 to 3 provided support for Covid-19 vaccine safety and effectiveness. It acted as the basis for global vaccine distribution. Due to the unprecedented acceleration of clinical trials, many questions remain revolving around appropriate vaccine regimes, the exact duration of protection, etc. Researchers need to investigate the long-term durability and effectiveness of Covid-19 vaccination and booster combinations, and the longevity of the produced antibody responses (Barouch, 2022).

References

Barouch, D.H. (2022). Covid-19 Vaccine - Immunity, Variants, Boosters. The New England Journal of Medicine, 387, 1011-1020. Retrieved December 17, 2022, from https://www.nejm.org/doi/full/10.1056/NEJMra2206573

Buchan, S. A., Chung, H., Brown, K. A., Austin, P. C., Fell, D. B., Gubbay, J. B., Nasreen, S., Schwartz, K. L., Sundaram, M. E., Tadrous, M., Wilson, K., Wilson, S. E., & Kwong, J. C. (2022). Estimated effectiveness of covid-19 vaccines against Omicron or delta symptomatic infection and severe outcomes. JAMA Network Open, 5(9). Retrieved December 29, 2022, from https://www.medrxiv.org/content/10.1101/2021.12.30.21268565v2.article-info

Chalkias, S., Eder, F., Essink, B., Khetan, S., Nestorova, B., Feng, J., Chen, X., Chang, Y., Zhou, H., Montefiori, D., Edwards, D. K., Girard, B., Pajon, R., Dutko, F. J., Leav, B., Walsh, S. R., Baden, L. R., Miller, J. M., & Das, R. (2022). Safety, immunogenicity and antibody persistence of a bivalent beta-containing booster vaccine against COVID-19: A phase 2/3 trial. Nature Medicine, 28(11), 2388–2397. Retrieved December 29, 2022, from https://www.nature.com/articles/s41591-022-02031-7

Chenchula, S., Karunakaran, P., Sharma, S., & Chavan, M. (2022). Current evidence on efficacy of COVID-19 Booster dose vaccination against the Omicron variant: A systematic review. Journal of Medical Virology, 94(7), 2969–2976. Retreived Decemeber 29, 2022, from https://onlinelibrary.wiley.com/doi/full/10.1002/jmv.27697

Fiolet, T., Kherabi, Y., MacDonald, C.-J., Ghosn, J., & Peiffer-Smadja, N. (2022). Comparing covid-19 vaccines for their characteristics, efficacy and effectiveness against SARS-COV-2 and variants of concern: A narrative review. Clinical Microbiology and Infection, 28(2), 202–221. Retreived December 29, 2022, from https://www.sciencedirect.com/science/article/pii/S1198743X21006042

Gilbert, P.B., Montefiori, D.C., McDermott, A.B., Fong Y., Benkeser, D., Deng, W., Zhou., H., Houchens, C., Martins, K., Jayashankar, L., Castellino, F., Flach, B., Lin, B.C., O'Connell, S., McDanal, C., Eaton, A., Sarzotto-Kelsoe, M., Lu, Y., Yu, C., Borate, B.,... Koup, R.A. (2021). Immune correlates analysis of mRNA-1273 COVID-19 vaccine efficacy clinical trial. Science, 375(6576), 43-50. Retrieved December 17, 2022, from https://www.science.org/doi/full/10.1126/science.abm3425

Hayawi, K., Shahriar, S., Serhani, M. A., Alashwal, H., & Masud, M. M. (2021). Vaccine versus variants (3Vs): Are the covid-19 vaccines effective against the variants? A systematic review. Vaccines, 9(11), 1305. Received December 29, 2022, from https://www.mdpi.com/2076-393X/9/11/1305

Kwok, H.F. (2021). Review of Covid-19 vaccine clinical trials - A puzzle with missing prieces International Journal of Biological Sciences, 17(16), 1461 - 1468. Retrieved December 17, 2022, from https://www.ncbi.nlm.nih.gov/pmc/articles/PMC8071768/

Lauring, A. S., Tenforde, M. W., Chappell, J. D., Gaglani, M., Ginde, A. A., McNeal, T., Ghamande, S., Douin, D. J., Talbot, H. K., Casey, J. D., Mohr, N. M., Zepeski, A., Shapiro, N. I., Gibbs, K. W., Files, D. C., Hager, D. N., Shehu, A., Prekker, M. E., Erickson, H. L., … Self, W. H. (2022). Clinical severity of, and effectiveness of mrna vaccines against, covid-19 from Omicron, Delta, and Alpha Sars-COV-2 variants in the United States: Prospective observational study. BMJ. Retrieved December 29, 2022, from https://www.bmj.com/content/376/bmj-2021-069761.long

Nasreen, S., Chung, H., He, S., Brown, K. A., Gubbay, J. B., Buchan, S. A., Fell, D. B., Austin, P. C., Schwartz, K. L., Sundaram, M. E., Calzavara, A., Chen, B., Tadrous, M., Wilson, K., Wilson, S. E., & Kwong, J. C. (2022). Effectiveness of covid-19 vaccines against symptomatic SARS-COV-2 infection and severe outcomes with variants of concern in Ontario. Nature Microbiology, 7(3), 379–385. Retrieved December 29, 2022, from https://www.nature.com/articles/s41564-021-01053-0

Ndwandwe, D., & Wiysonge, C. (2021). Covid-19 vaccines. Current Opinion in Immunology, 71, 111 - 116. Retrieved December 17, 2022, from https://www.sciencedirect.com/science/article/pii/S095279152100090X

Pormohammad, A., Zarei, M., Ghorbani, S., Mohammadi, M., Razizadeh, M.H., Turner, D.L, & Turner, R.J. (2021). Efficacy and Safety of COVID-19 Vaccines: A Systematic Review and Meta-Analysis of Randomized Clinical Trials. Vaccines, 9(5). Retrieved December 17, 2022, from https://www.mdpi.com/2076-393X/9/5/467

Rahman, M., Masum, H.U., Wajed, S., & Talukder, A. (2022). A comprehensive review on COVID-19 vaccines: development, effectiveness, adverse effects, distribution and challenges. Virus Disease, 33, 1-22. Retrieved December 17, 2022, from https://link.springer.com/article/10.1007/s13337-022-00755-1

Vetter, V. Denizer, G. Leonard, R. F., Krishnan, J., & Shapiro, M. (2018). Understanding modern-day vaccines: what you need to know. Annals of Medicine, 50(2), 110-120. Retrieved from December 17, 2022, from https://www.tandfonline.com/doi/full/10.1080/07853890.2017.1407035

Mental Health and Psychiatric Impact

By Jessica Henschel

In the past two years, everyone across the globe has felt the weight of the novel coronavirus pandemic, COVID-19. Drastic public health measures were taken to combat the spread of the virus and reduce the strain on hospitals. Many countries opted for intensive lockdowns and restrictions that severely limited social interaction to members of a household and only allowed citizens to leave their homes for groceries and medical appointments. Despite the beneficial effect these lockdowns had for stopping the spread of the virus, it has come at the cost of mental and social health. This left us all wondering: what impacts has COVID-19 had on the mental health of North Americans? The fist half of this chapter will explore how psychiatric and mental health care have traditionally responses to pandemics in the past. The second half will review the variety of public health orders employed in Canada and the USA in response to the COVID-19 pandemic and how these restrictions have impacted mental health.

Psychiatric Responses: The History of Pandemic Responses

Although COVID-19 restrictions have been majorly dropped across Canada and the USA, there have been significant psychiatric and psychological impacts that will continue to affect public health. Psychiatry has been at the forefront of pandemic responses throughout history, even if it was not always recognized or refered to as such. Mental health disorders have been identified in various populations across the world for centuries and their prevalence seems to have remained. However, it is evident that mental illnesses have been exacerbated and increased in prevalence during pandemics and as a whole. This section will review the pandemic known as the Black Death and its effect on mental health and the development of psychiatry.

Prior to modern psychiatry/psychology in the Westen world, disorders such as schizophrenia, bipolar disoroder, and major depressive disorder were all refered to as "insanity" or "madness" (Benedictow, 2005). In the last 200 years, the rate of mental health disorders have increased at least fivehold. There is no specific smoking gun for this increase, but it is thought that mental health has become more acceptable to speak about in the last few decades. Prior to the 21st century, it was common for any individual exhibiting mental disorders to be shunned and unwillingly thrown into state hospitals or asylums. Asylums began popping up in Europe as early as the Medieval period and were in use during one of the most well-known world pandemics- the Black Death (Benedictow, 2005). Often, asylums served as makeshift prisons where anyone who did not fit social norms was sent. These buildings were poorly kept and overrun with people, disease, and flith. As these people began to be viewed as those with severe psychiatric conditions instead of criminals or deviants, mental health has become increasingly talked about and studied.

Various shifts in public health and the view on mental health occurred during the time of the infamous Black Death in Europe. Between 1347-1351, the epidemic commonly known as the Black Death swept across Europe, killing more than 20 million people (25% of the population) (Lindemann, 2010). Unlike during the Plague of Athens, there were many sources and firsthand accounts in which information was recorded about the plague, informing us of the atmosphere in Europe when the disease first appeared. As Lindemann (2010) details, these sources informed us that the plague was believed to have entered the continent from 12 ships from the Black Sea docked at the Sicilian port of Messina. Most sailors aboard the ship were either dead or severely ill and covered in black boils that oozed blood and pus. Despite the Sicilian authorities' attempt to rid the harbor of all the so called "death ships", the plague had already begun to spread (History.com, 2020).

The Black Death devastated Europe and Asia during the first 5 years after its arrival in Messina but continued to last for 400 years after until it finally vanished (Lindemann, 2010). It was thought to be spread by trading ships, making it rampant and infectious to every major city with a port. It was later hypothesized that the true carrier of the plague on these trading ships was rats, as they were carriers of the bacteria. Scientists and historians in modern times have suggested that the rats were carrying the Yersinia Pestis bacteria (bubonic plague), which would be hosted by fleas on the rodents (Lindemann, 2010). Once the carrier rat died, the fleas would hop onto humans and bite them, thus infecting them with the bacteria. Once a human was bitten, the site swelled to form a painful and large bubo, most often in the groin, thigh, armpit, or neck. Victims began to show signs 3-5 days after they were initially infected, usually dying another 3-5 days later (80% of the cases) (Benedictow, 2005).

Despite many advertisements for remedies and healers that could cure the plague, mass death still occurred late into 1351 and beyond (Lindemann, 2010). As no medicines or doctors seemed to help, many came up with alternative methods of explaining and living through the plague. There were various explanations and responses to the Black Death, with the most prominent being religion. The common understanding in the predominantly Christian areas was that God was punishing the sinners with the great dying and saving the believers (Lindemann, 2010). Due to the lack of psychiatric knowledge at this time, the churches responded by sending out prayers to be recited by the public and held large masses to pray for reprieve. Due to this belief, they thought that the only way to be saved was to gain God's forgiveness and purge their communities of heretics (History.com, 2020). Another well known response to the plague were a group of individuals called flagellants. Upperclassmen were known to join processions of traveling flagellants, who engaged in public displays of punishment (History.com, 2020). They would beat themselves and others in with leather straps studded with metal in public areas for 33 and ½ days. Townspeople would watch as these men inflicted pain on one another 3 times a day. This movement was targeted towards repenting for their sins and seeking penance. Modern day scholars regard flagellants as those acting out non-suicidal self-harm as a response to the extreme distress and terror from the Black Death. Often self-harm is associated with mental health disorders such as depression, boderline personality disorder, and bipolar disorder- all of which are exacerbated by pandemic conditions.

As Lindemann (2010) emphasizes in her review of medicine and early European society, it is still unknown how great of an impact the Black Death had on European society as a whole. People began to migrate to less populated areas, resulting in clearing of forests and reliance on animal husbandry (Benedictow, 2005). One of the most significant societal and scientific changes that occurred due to the plague was the increase in public health measures. The loss of life was significant and medical and public health orders began to develop out of pure necessity to limit the spread (Lindemann, 2010). Officials began to set up quarantines, where the infamous plague doctors in their beak-like masks would visit to assess symptoms. Subsequently, fumigation was arranged for people in quarantine and for their possessions in an attempt to clear the miasma or "bad air", which was how the plague was thought to be spread (Lindemann, 2010). Additionally, special hospitals were built for the further isolation of patients and large gatherings were prohibited. However, this did not address the adverse psychological effects the plague had. People were demoralized, terrified, and suffered enormous losses- in terms of lives and livelihoods. Some responded by acting out and disobyeing public health orders, partying and engaging in debauchery. Others were too scared and beat down to leave their homes.

The emergence of public asylums was in part a response to viral outbreaks that were unexplainable with the limited technology of the time. In London, one of the most notorious mental health hospitals was the Priory of Saint Mary of Bethelehem: better known as Bedlam asylum (Porter, 2006). Although Bedlam was founded in 1247, it began hosting mentally ill patients in the years just after the Black Death virus had started to die down and transmission was less likely. Unfortunately, the conditions were horrific and extremely inhumane; patients were kept in prison cells and it was recorded that the hospital had manacles, chains, stocks, and areas for solitary confinement. Not much is known about the treatment of patients during this time, but more and more individuals started being incarcerated (indefinitely) at Bedlam as the years went on. These individuals were not only those with mental illness, but those who were seen as social deviants or abormal from the rest of society (Porter, 2006).

Present Day Mental Health Impacts of COVID-19

When the first case of COVID-19 was recorded in Canada on January 25, 2020, public health officials across the country began to prepare for the worst (Bronca, 2020). Patient zero — a man who had recently returned from the outbreak epicentre, Wuhan, China to Toronto — was immediately put into quarantine and major airports began introducing extensive screening measures. In February 2020, China's numbers were rapidly increasing, and Canada began seeing a rise in travel-related cases. The exponential growth of cases in March were enough to see almost every province declare a state of emergency (Bronca, 2020). All retail, gyms, recreation centres, restaurants, casinos/banquet halls, and concert venues were shut down. In the height of the first wave (March-June 2020), citizens were only allowed to go out for groceries and medical visits. Social gathering were prohibited, and physical contact was limited to household members only (Karstens-Smith, 2020). Due to the public health acts, it was illegal to break the lockdown rules and would result in fines or jail time. This came as an understandable shock and brought about an innate sense of loss as people across the world were laid off, forced to close their businesses, could not attend school, and most significantly, not see their friends or family. As Karstens-Smith (2020) reported for Global News, the new restrictions were met with outrage and tremendous fear. Unfortunately, none of us predicted that these conditions would still be occurring in sporadic waves two years later.

Due to the loss of lives and livelihoods during the COVID-19 pandemic, the mental health of Canadians has been negatively impacted. Mental Health Research Canada (MHRC) released that there had been a significant increase of adult Canadians reporting depression (22%) and anxiety (20%) disorder diagnoses (Flanagan, 2021). These numbers were the highest rates MHRC had ever measured, with 6% of Canadians (1.8 million) having all 4 negative indicators that polling uses to track mental health: high anxiety and depression, moderate to severe mental health symptoms, low management of stress, and low resiliency (MHRC, 2021). In addition to MHRC's survey, the Canadian Mental Health Association (CMHA) partnered with University of British Columbia to provide a detailed study of the effects of COVID-19 on psychological health in Canada (The University of British Columbia [UBC], 2020). They reported that the second wave of the pandemic brought about increased levels of stress and anxiety. About 71% of Canadians stated they were worried about the continued waves of the pandemic, with 58% being worried about the mortality of their loved ones (UBC, 2020).

Of greatest concern to mental health service providers and the general population was the sharp increase in suicidality since the COVID-19 pandemic began. With 1 in 10 Canadians experiencing feelings or thoughts of suicide, public health researchers noticed the severe psychological effects of the virus and restrictions (UBC, 2020). Suicidal ideation was even stronger amongst already marginalized groups, such as those who identify at LGBTQIA2S+, individuals with pre-existing mental illness, disabilities, and Indigenous peoples. Mental health and addictions researcher at UBC, Emily Jenkins stated, "we are seeing a direct relationship between social stressors and declining mental health. Particularly concerning are the levels of suicidal thinking and self-harm, which have increased exponentially since before the pandemic [...]" (UBC, 2020). One of the primary stressors Jenkins referred to is the financial strain from the pandemic. More than a third of Canadians were concerned about finances, with the majority of individuals being parents with small children. Many parents are burdened with providing for their families and being able to put food on the table while they have lost their jobs and stability due to lockdowns.

We are seeing similar impacts in the United States of America (USA). Similarly to Canada, when the COVID-19 pandemic broke in the USA, it drastically affected economics, politics, and health care. The unemployment rate in 2020 went from 3.5% in March to 14.7% by April, with some sources even suggesting 19.7% (Blanchflower & Bryson, 2022). Since then, mental health impacts have been extreme, especially for young people in America. Mental Health America (MHA) reported that 35% of individuals screened reported depression and 20% had anxiety. However, anxiety rates skyrocketed to 80% by September 2020 (Blanchflower & Bryson, 2022). There is evidence that the impacts of COVID-19 as a whole has increased not only depression, anxiety, and stress, but has amplified even more serious mental health disorders (World Health Organization [WHO], 2022). Specifically, mental health professionals are seeing more cases of post-traumatic stress disorder (PTSD) and trauma responses from the pandemic. The abrupt response and effects of the pandemic led to the loss of livelihoods and the death of loved ones, leading to mass trauma responses from Americans. This was exacerbated by social isolation and disconnectedness from others, especially in marginalized communities that rely on the support of others in the same group (such as BIPOC and LGBTQIA2S+ communities).

With the mental health crisis spurred by the pandemic, Americans are also facing increased rates of suicide, overdoses, domestic abuse, and alcohol poisoning (Panchal et al., 2021). After lockdowns began, EMS and hospital staff reported a upward incline of opioid overdoses and gun violence, as well as suicide attempts and self-harm. It was found that many of the individuals attempting suicide were young adults and adolescents (Panchal et al., 2021). It is speculated that young adults (aged 18-24) were hit the hardest by the pandemic due to university/college closures, loss of socialization at a crucial developmental time, job loss, and loss of income. Young adults are already at risk of poor mental health outcomes and suicide, and the COVID-19 pandemic has made this even worse. Unfortunately, these circumstances have led to increased suicidal ideation and suicide attempts in young people (as well as older individuals). After a two year decline in suicide rates, the pandemic spurred suicide rates to rise and 2021 saw almost 48,000 deaths by suicide. The result was one death every 11 minutes in the USA from suicide or 14 deaths for every 100,000 people (McPhillips, 2022). Substance use disorder and overdoses have also been going up since March 2020 as a way for individuals to cope with the loss of normlacy and in response to stress. Drug and alcohol induced deaths were particularly pronounced between March-May 2020, but these disorders have persisted into 2022 (Panchal et al., 2021).

A study by McIntyre and Lee (2020) used macroeconomic indicators (primarily unemployment) from the COVID-19 pandemic to predict suicide statistics in Canada. Based on the suicide and unemployment rates from previous years, McIntyre and Lee (2020) calculated various mortality projections for 2021. If unemployment continues to rise in the country at the same rate as 2020, the projected suicide rate would increase to 13.6% in 2021. This would result in an additional 2,114 suicides across the country. Devastatingly, these deaths could have been preventable if Canada and the USA had handled the pandemic response in a different manner in March 2020. Therefore, these numbers merit the question: are lockdowns worth it? The economic costs of lockdowns seem to far outweigh the health benefits in some cross-country studies (Thom, 2021). Seemingly the loss of life from suicide may be preventable as well. Despite the mental health crisis in the world, lockdowns may be the only feasible option in preventing strain on healthcare systems and hospitals. Both Canada and the USA have suffered from immense strain on their health care systems- burnout in hospital staff, nurses, and doctors, as well as a lack of ICU and inpatient hospital beds. There is still a lack of resources and funding for those who are ill, with COVID-19 patients taking up the majority of beds and ventilators. Lockdowns were the only feasible way to reduce this strain, but it has had its consequences in other health areas. Further research and policy work must be done to ensure that the physical and mental health of citizens are being protected, while still doing what is in the best interest of the economy.

Resources for Mental Health

Due to the current mental health crisis in Canada, there are several supports that have been made available. The Mental Health Commission of Canada has created a program called Wellness Together Canada, which provides 24/7 mental health support to Canadians (Mental Health Commission of Canada, 2021). Wellness Together Canada provides various services: access to free, live counselling through either text (SMS) or phone, addiction help, search for psychological resources in your area, and self-care information/activities. Every province and territory has set up their own government supports as well, such as Here to Help British Columbia, Help in Tough Times Alberta, Care for Your Mental Health Manitoba, COVID-19 Support for People Ontario (Mental Health Commission of Canada, 2021). If you are experiencing suicidal ideation or struggling with your psychological well-being, please contact your nearest government support or call 911 for a mental health emergency.

MHA has created a website in response to the pandemic and to help Americans with their mental health (MHA, 2022). This website provides various resources for emergency mental health situations and for long-term recovery. It is tailored to different populations as well: frontline workers, caregivers, parents, BIPOC people, and LGBTQA+ individuals. It provides wellness and coping skills, such as resources on re-entering college, accessing therapy, managing loneliness, stress, worry, and negative self-talk. MHA's website also offers a variety

81

of mental health screening tools to assist individuals in determining if their mental health is taking a decline and how to access help. Online testing is a great way to keep tabs on your mental health and to monitor if the results are abnormal for us. It also helos us to identify what exactly is going on and how we can address it. These tests are statistically validated and can monitor for: depression, anxiety, psychosis, eating disorders, bipolar disorder, addictions, PTSD, and ADHD (MHA, 2022). If you find you are an American and experiencing suicidal ideation or struggling with your mental health, please consider texting or calling 988 or using the chat box at 988lifeline.org/chat.

References

Benedictow, O.J. (2005, March 3). The black death: The greatest catastrophe ever. History Today. https://www.historytoday.com/archive/black-death-greatest-catastrophe-ever

Blanchflower, D.G., & Bryson, A. (2022). Covid and mental health in America. PLoS One, 17(7). https://doi.org/10.1371/journal.pone.0269855

Bronca, T. (2020, April 8). COVID-19: A Canadian timeline. Canadian Healthcare Network. https://www.canadian-healthcarenetwork.ca/covid-19-a-canadian-timeline

Karstens-Smith, B. (2020, December 28). A timeline of COVID-19 in Alberta. Global News. https://globalnews.ca/news/7538547/covid-19-alberta-health-timeline/

Flanagan, R. (2021, January 14). Canadians reporting more anxiety and depression than ever before, poll finds. CTV News. https://www.ctvnews.ca/health/coronavirus/canadians-repor ting-more-anxiety-and-dep ression-than-ever-before-poll-finds-1.5266911

History.com. (2020, July 6). Black death. https://www.history.com/topics/middle-ages/blackdeath

Lindemann, M. (2010). Medicine and society in early modern Europe (2nd ed.). Cambridge University Press.

Mental Health Commission of Canada. (2021). Government of Canada COVID-19 resources. https://www.mentalhealthcommission.ca/English/government-canada-covid-19-resource

Mental Health America. (2022). COVID-19. https://mhanational.org/covid19

Mental Health Research Canada [MHRC]. Mental health in crisis: How COVID-19 is impacting Canadians. https://www.mhrc.ca/national-poll-covid/findings-of-poll-5

McIntyre, R.S., & Lee, Y. (2020). Projected increases in suicide in Canada as a consequence of COVID-19. Psychiatric Research, 290, 104-113. https://doi.org/10.1016/j.psychres.2020.113104

McPhillips, D. (2022). US suicide rates rose in 2021, reversing two years of decline. CNN. https://www.cnn.com/2022/09/30/health/suicide-deaths-2021/index.html

Panchal, N., Kamal, R., Cox, C., & Garfield, R. (2021). The implications of COVID-19 for mental health and substance use. KFF. https://www.kff.org/coronavirus-covid-19/issue-brief/the-implications-of-covid-19-for-m ental-health-and-substance-use/

Porter, R. (2006). Madmen: A social history of madhouses, mad doctors, and lunatics. Tempus.

Thom, H. (2021, February 10). Lockdown critics are sure
    the costs outweigh the health benefits,but they're wrong.
    Phys.org. https://phys.org/news/2021-02-lockdown-critics-
    outweigh-health-benefits.html

The University of British Columbia. (2020, December 3).
    New national survey finds Canadians' mental health erod-
    ing as pandemic continues. https://www.med.ubc.ca/news/
    new-national-survey-finds-canadians-mental-health-erod-
    ing-as-pandemic-continues/

World Health Organization. (2022, June 16). The impact
    of COVID-19 on mental health cannot be made light of.
    https://www.who.int/news-room/feature-stories/detail/
    the-impact-of-covid-19-on-mental-health-cannot-be-made-
    light-of

Life Outside the Pandemic: Health Promotion

By Massa Mohamed Ali

The varying effects of the pandemic on individuals leaves one wondering - why did some people get more affected by the virus than others? Why is it that even those of the same age, gender, and general health conditions still react to COVID-19 differently? How did patients, even the relatively healthy ones, develop irreversible symptoms? These questions may arise from the frustration felt after dealing with the damage done by COVID-19. We have repeatedly seen that people react differently to COVID-19. Its clinical spectrum ranges from asymptomatic to severe illness and death. Globally, around 40% of those infected with COVID-19 were asymptomatic (Ma et al., 2021). Yet, there were over 6 million total deaths from COVID-19 worldwide (World Health Organization, 2022). So what does one's reaction to the virus depend on? Why are some individuals able to fight it off more easily than others? And how can we be of those who fight it off quickly and even silently?

The answers to these questions depend on a number of factors, one largely being our immune system. While many believe that the pandemic was inevitable, we can do our part by being prepared for it and for future pandemics like it. We can protect ourselves, support our immune system, and increase our fitness to ensure the best response to illness. This can be

85

achieved through health promotion. Health promotion is a broad term that refers to the actions taken to improve the health of individuals and communities. It involves a holistic approach that considers the social, economic, and environmental factors that influence health (Kumar & Preetha, 2012). Prioritizing health promotion is essential to ensure that individuals and communities are better prepared to prevent and manage future public health crises. Health promotion after the COVID-19 pandemic should also focus on the prevention of future pandemics. This can include measures such as vaccination, infection control measures, and improving public health infrastructure. The COVID-19 pandemic has highlighted the importance of preparedness and the need for strong public health systems to prevent and manage public health crises.

Health promotion focuses on the lifestyle-related issues that contribute to chronic disease. As one thing leads to another, environmental risk factors such as financial issues, unstable relationships, and low education can lead to behavioural risk factors such as physical inactivity, an unhealthy diet, or tobacco use, which then lead to biological risk factors like high blood pressure, high blood glucose, and increased inflammation in the body (Kumar & Preetha, 2012). All this can eventually cause heart disease, strokes, cancer, and more, which can affect one's immunity to other illnesses like COVID-19 (Kumar & Preetha, 2012).

Stress

The various environmental risk factors all have one outcome in common - stress. Stress has been shown to weaken the body's immune system by increasing the amount of cortisol in the body. With more cortisol circulating in the blood, more pro-inflammatory cytokines are released. Cytokines are small proteins produced by immune system cells to fight off infection and heal injuries. Examples include interleukin-1 (IL-1), interleukin-6 (IL-6), and tumor necrosis factor-alpha (TNF-alpha) (Turner et al., 2014). This induces a state of inflammation in the body and makes it more susceptible to infection and disease.

Stress also directly affects the brain, further impacting one's ability to cope and creating a dangerous cycle. The more stress, the worse the anxiety and ability to handle stress, leading to even more stress. During acute stress, the hypothalamic-pituitary-adrenal (HPA) axis is stimulated by pro-inflammatory cytokines to release glucocorticoids (GCs). These GCs then inhibit pro-inflammatory activity, which is beneficial in the short term. When the stress becomes chronic over the long-term, the prolonged exposure to GCs makes the peripheral immune system less sensitive to anti-inflammatory signals in the brain, causing a state of inflammation. This pro-inflammatory state affects the activity and connections of the dorsal anterior cingulate cortex (dACC) and the insula, which modulate the switch between internally and externally directed cognition. This disruption often underlies disorders like anxiety, as it leaves one unable to detach their thoughts from reality. One of the main reasons for the disruption is the increased release of glutamate by glial cells in response to the immune system's activation, which leads to synaptic dysfunction and apoptosis in several brain regions (Brenmer, 2006).

The pandemic has caused a significant amount of stress and anxiety for many people, and it is important to provide support and resources to help individuals cope with these challenges. There is strong evidence to support the benefits of mental health promotion in improving health outcomes. A systematic review of over 200 studies found that interventions aimed at promoting mental health are associated with a reduced risk of all-cause mortality, cardiovascular disease, and certain types of cancer (Patel et al., 2007). Another systematic review found that mental health promotion interventions are also associated

87

with a reduced risk of cognitive decline and dementia (Baumann, Matzek, & Schönknecht, 2017). In addition to these health benefits, mental health promotion can have economic benefits. A study conducted by the World Health Organization found that investing in mental health promotion can result in cost savings for both individuals and governments (World Health Organization, 2010).

To promote mental health after the COVID-19 pandemic, it is important to consider the barriers that may prevent individuals from seeking support and resources for their mental health. These barriers can include stigma, lack of access to mental health services, and lack of awareness of available resources (World Health Organization, 2020). To overcome these barriers, it may be necessary to implement policies and programs that reduce stigma, increase access to mental health services, and increase awareness of available resources (World Health Organization, 2020). Examples of these types of policies and programs include training for primary care providers in mental health screening and referral, providing mental health education and support to schools and workplaces, and providing access to online resources and hotlines for mental health support.

Exercise

One important aspect of health promotion and reducing stress after the COVID-19 pandemic is the promotion of physical activity. This healthy form of physical pressure on the body helps it fight inflammation by acting as an acute stressor, only releasing cortisol for a short amount of time and inhibiting pro-inflammatory activity. Physical activity has numerous health benefits, including reducing the risk of chronic diseases such as obesity, heart disease, and diabetes. It has also been shown to improve mental health, reduce stress and anxiety, and improve overall quality of life (World Health Organization,

2020). The COVID-19 pandemic has led to an increase in sedentary behaviour as many people have been confined to their homes. This has made it even more important for people to prioritize physical activity in their daily lives.

There is strong evidence to support the benefits of physical activity in improving health outcomes. A systematic review of over 400 studies found that physical activity is associated with a reduced risk of all-cause mortality, cardiovascular disease, breast and colon cancer, and type 2 diabetes (Warburton, Nicol, & Bredin, 2006). Another systematic review found that physical activity is also associated with a reduced risk of cognitive decline and dementia (Lautenschlager, Cox, Flicker, Foster, & van Bockxmeer, 2008). Moreover, researchers at the London School, Harvard, and Stanford compared exercise to pharmaceutical interventions and found that exercise often worked just as well as drugs for the prevention of diabetes and the treatment of heart disease and stroke. Exercise has therefore been shown to significantly extend one's lifespan. In addition to these health benefits, physical activity has also been shown to have economic benefits. A study conducted by the World Health Organization found that investing in physical activity promotion can result in cost savings for both individuals and governments (World Health Organization, 2010).

The reduced risk of illness with better fitness is because consistent physical activity reduces fat tissue and increases the production of anti-inflammatory cytokines (Woods et al., 2012). According to research done at the University of California San Diego School of Medicine, just one session of moderate exercise is sufficient for stimulating the anti-inflammatory effects of the immune system (Dimitrov et al., 2017). Researchers showed that 20-30 minutes of moderate exercise, including fast walking, resulted in a 5% decrease in the number of stimulated immune cells producing TNF, a cytokine responsible for regulating local and systemic inflammation (Dimitrov et al., 2017). In fact, daily exercise is so important that not walking an hour a day is considered 'high-risk' behaviour according to Dr. Michael Greger, a world-renowned American physician, author, and public speaker on public health issues.

To promote physical activity after the COVID-19 pandemic, it is important to consider the barriers that may prevent individuals from being physically active. These barriers can include time constraints, lack of access to facilities, and lack of motivation (World Health Organization, 2020). To overcome these barriers, it may be necessary to implement policies and programs that increase access to physical activity opportunities and provide support and encouragement to individuals to be physically active. Examples of these types of policies and programs include building or improving walking and cycling infrastructure, promoting the use of public transportation, and providing incentives for individuals to be physically active (World Health Organization, 2020). This can include promoting standing desks, treadmills at home, cycling chairs, or consistent move breaks at work.

Nutrition

In addition to promoting physical activity, it is also important to focus on the promotion of a healthy diet after the COVID-19 pandemic. A healthy diet is essential for maintaining good health and preventing chronic diseases. It is defined as a diet that is rich in fruits, vegetables, and whole grains and low in unhealthy fats, added sugars, and sodium (World Health Organization, 2020). The COVID-19 pandemic has led to an increase in the consumption of unhealthy foods, such as processed and fast foods, as people have turned to comfort eating during times of stress. This has made it even more important for people to focus on consuming a healthy diet.

There is strong evidence to support the benefits of a healthy diet in improving health outcomes. A systematic review of over 100 studies found that a healthy diet is associated with a reduced risk of all-cause mortality, cardiovascular disease, and certain types of cancer (World Cancer Research Fund/ American Institute for Cancer Research, 2018). Another systematic review found that a healthy diet is also associated with a reduced risk of cognitive decline and dementia (Liu, Ming, & Fung, 2017). According to the World Health Organization (2020), a healthy diet should include a variety of foods from all food groups, including:

• "Fruits and vegetables: at least 400 grams per day

• Whole grains: at least three servings per day

• Legumes: at least three servings per week

• Nuts and seeds: at least one serving per day

• Lean protein sources: such as poultry, fish, and legumes

• Low-fat dairy products: such as milk, cheese, and yogurt"

Several factors may contribute to unhealthy eating habits. These factors can include the availability and accessibility of healthy food options, the influence of marketing and advertising, and cultural and social norms (World Health Organization, 2020). People may find themselves resorting to food for comfort instead of health, which can lead to disordered eating and addiction to specific foods. To address these factors, it may be necessary to implement policies and programs that increase the availability and accessibility of healthy food options, limit the marketing and advertising of unhealthy foods, and promote cultural and social norms that support healthy eating habits (World Health Organization, 2020). Raising awareness about the dangers of uncontrolled eating and food addiction is fundamental, especially for children and adolescents.

More specifically, the overconsumption of sugar in our society has been a leading cause of many chronic illnesses and diseases, further weakening the immune system. Studies have shown that eating a high sugar, high carbohydrate processed diet makes the body less resistant to bacteria, viruses, and parasites (Livingston, 2020). One study showed that it takes about 75 grams of sugar to weaken the immune system, which once affected does not return to normal until about 5 hours after sugar consumption (Livingston, 2020). This means that even with consistent physical exercise, stress reduction, and proper sleep, eating a diet high in sugar and refined carbohydrates can compromise the immune system. Dr. Michael Greger has a "Daily Dozen" list, where he recommends the best anti-inflammatory foods to incorporate in our diet on a daily basis. On his website nutritionfacts.org he writes:

"Each day, I recommend a minimum of three servings of beans (legumes), one serving of berries, three servings of other fruits, one serving of cruciferous vegetables, two servings of greens, two servings of other veggies, one serving of flaxseeds, one serving of nuts and seeds, one serving of herbs and spices, three servings of whole grains, five servings of beverages, and one serving of exercise (90 minutes at moderate intensity or 40 minutes of vigorous activity).

This may sound like a lot of boxes to check, but it's easy to knock off several at once. With one peanut butter and banana sandwich, you've just checked off four boxes. Sit down to a big salad of two cups of spinach, a handful of arugula, a handful of walnuts, a half cup of chickpeas, a half cup of red bell pepper, and a small tomato, and seven boxes can be ticked in one dish. Sprinkle on your flax, add a handful of goji berries, and enjoy it with a glass of water and fruit for dessert, and you could wipe out nearly half your daily check boxes in one meal. And then if you ate it on a treadmill…(kidding!)." ("Daily Dozen," n.d.)

Sleep

Another fundamental aspect of health promotion is ensuring that individuals get quality sleep. Sleep affects both the innate and adaptive parts of the immune system. Just as sleep is important for long-term learning and memory consolidation, studies show that it strengthens immune memory as well (Suni, 2022). During sleep, several immune system components interact to remember to react to and recognize pathogens encountered throughout the day. Research has also shown that cytokines are produced during sleep to generate inflammation in the body (Suni, 2022). This nighttime activity strengthens adaptive immunity during a time when minimal energy is needed for the rest of the body. In addition, melatonin, the sleep-promoting hormone produced at night, is able to counteract the stress that comes from inflammation during sleep (Suni, 2022). In this way, sleep maintains the delicate balance of the immune system and promotes the body's self-healing processes. Studies have also found that those who do not get at least 7 hours of quality sleep after receiving a hepatitis or swine flu vaccine had a significantly weaker immune response, suggesting that their bodies may not have had enough time to develop the necessary immunological memory in response to the vaccine.

Sleep deprivation is known to be linked to a number of health issues, mainly for its damaging effects on the immune system. In those who sleep enough, the inflammation they experience recedes back to normal as they wake up. In those who lack quality sleep, this inflammation persists. The body becomes stressed and inflamed, therefore disrupting the immune system and leaving cytokines circling throughout the body. This makes one more susceptible to heart disease, diabetes, strokes, cancer, and Alzheimer's disease.

Additionally, sleep supports the activity of T cells responsible for fighting intracellular pathogens (Pratt, 2019). Researchers at the University of Tubingen found that after one night, participants who slept had higher levels of integrin activation in their T cells than those who did not sleep that night (Dimitrov et al., 2019). Integrin activation is essential for T cells to kill virus-infected cells or cancer cells (Dimitrov et al., 2019). Therefore, an adequate amount of sleep has the potential to improve T cell functioning to support optimal functioning of the immune system. The Centres for Disease Control and Prevention recommends practicing good sleep habits to get a good night's sleep. These include having a consistent sleeping and waking time, including on weekends; making sure one's bedroom is quiet, dark, and at a comfortable temperature; removing electronic devices from the bedroom; avoiding large meals at least 3 hours before bedtime; and being physically active throughout the day ("Tips for Better Sleep," 2022).

Supplements

With a healthy balanced diet, exercise, proper sleep, and reduced stress, the immune system can function optimally and fight off viruses like COVID-19 a lot quicker and faster. While some may believe that boosting our immune system with vitamin supplements may be helpful, the process is not this simple. Taking vitamin supplements is only beneficial when one has a true deficiency. Some studies show that some supplements, when not medically needed, may have negative effects on the immune system. Dr. Jen Gunter, a Canadian-American gynecologist and a specialist in chronic medicine and women's health, suggests thinking of the immune system as a garden to support, not as a muscle to strengthen. She explains that the immune system has many components, just like a garden has many kinds of plants with each plant needing just the right amount of sun, shade, and water. Too much sun might help one plant but destroy another, she says. Therefore, the immune system maintains a balance between all components and too much of one element may destroy this balance, leading to autoimmune diseases or failure to fight infection. If one is getting all the essential nutrients from their diet, taking more of one nutrient will not help the immune system. If taking more than one needs, the body can just get rid of the excess as waste.

Vaccines

Furthermore, Dr. Gunter says the one thing repeatedly shown to be beneficial for the immune system is vaccines. They contain an inactive or weakened part of a pathogen that includes its antigen nametag. This triggers the immune system to make the necessary antibodies without getting sick and overwhelmed. Vaccines provide the system with exactly what it needs to remember the appropriate way to respond to a pathogen when it enters the body. To prevent future pandemics, it is important to invest in the development and distribution of vaccines, as well as in the infrastructure needed to deliver vaccines to individuals. There is strong evidence to support the effectiveness of vaccination in preventing infectious diseases and reducing morbidity and mortality (World Health Organization, 2020). In addition to vaccination, control measures to control infection must be continuously implemented, such as hand hygiene and sanitization areas, to reduce the transmission of infectious diseases. Finally, investing in public health infrastructure, including surveillance systems, laboratory capacity, and healthcare facilities, ensures that public health systems are prepared to respond to future public health crises.

References

Baumann, N., Matzek, W., & Schönknecht, P. (2017). Promoting mental health in old age: A systematic review of randomized controlled trials. Aging & Mental Health, 21(6), 634-646.

Bremner J. D. (2006). Traumatic stress: effects on the brain. Dialogues in clinical neuroscience, 8(4), 445–461. https://doi.org/10.31887/DCNS.2006.8.4/jbremner

Centers for Disease Control and Prevention. (2022, September 13). Tips for better sleep. Centers for Disease Control and Prevention. Retrieved December 31, 2022, from https://www.cdc.gov/sleep/about_sleep/sleep_hygiene.html

Daily Dozen. Nutritionfacts.org. (n.d.). Retrieved December 31, 2022, from https://nutritionfacts.org/topics/daily-dozen/

Dimitrov, S., Lange, T., Gouttefangeas, C., Jensen, A.T.R., Szczepanski, M., Lehnnolz, J., Soekadar, S., Rammensee, H., Born, J., Besedovsky, L.; Gas-coupled receptor signaling and sleep regulate integrin activation of human antigen-specific T cells. J Exp Med 4 March 2019; 216 (3): 517–526. doi: https://doi.org/10.1084/jem.20181169

Flaherty, S. (2021, October 13). Study confirms kids as spreaders of covid-19 and emerging variants. Harvard Gazette. Retrieved December 31, 2022, from https://news.harvard.edu/gazette/story/2021/10/study-confirms-kids-as-spreaders-of-covid-19-and-emerging-variants/

Kumar, S., & Preetha, G. (2012). Health promotion: an effective tool for global health. Indian journal of community medicine : official publication of Indian Association of Preventive & Social Medicine, 37(1), 5–12. https://doi.org/10.4103/0970-0218.94009

Lautenschlager, N. T., Cox, K. L., Flicker, L., Foster, J. K., & van Bockxmeer, F. M. (2008). Physical activity for cognitive and mental health in older people: A review. Journal of Aging and Physical Activity, 16(3), 240-259.

Liu, C., Ming, X., & Fung, H. H. (2017). A healthy diet and cognitive function in middle-aged and older individuals: A systematic review. Journal of Gerontology: Series A, 72(9), 1292-1305.

Patel, V., Flisher, A. J., Hetrick, S., & McGorry, P. (2007). Mental health of young people: A global public-health challenge. The Lancet, 369(9569), 1302-1313.

Pratt, E. (2019, February 21). Sleep and immune system. Healthline. Retrieved December 31, 2022, from https://www.healthline.com/health-news/how-sleep-bolsters-your-immune-system

Qiuyue Ma, P. D. (2021, December 14). Global percentage of asymptomatic SARS-COV-2 infections. JAMA Network Open. Retrieved December 31, 2022, from https://jamanetwork.com/journals/jamanetworkopen/fullarticle/2787098

Sleep & immunity: Can a lack of sleep make you sick? Sleep Foundation. (2022, April 22). Retrieved December 31, 2022, from https://www.sleepfoundation.org/physical-health/how-sleep-affects-immunity

Turner, M. D., Nedjai, B., Hurst, T., & Pennington, D. J. (2014). Cytokines and chemokines: At the crossroads of cell signalling and inflammatory disease. Biochimica et biophysica acta, 1843(11), 2563–2582. https://doi.org/10.1016/j.bbamcr.2014.05.014

Warburton, D. E., Nicol, C. W., & Bredin, S. S. (2006). Health benefits of physical activity: The evidence. Canadian Medical Association Journal, 174(6), 801-809.

Woods, J. A., Wilund, K. R., Martin, S. A., & Kistler, B. M. (2012). Exercise, inflammation and aging. Aging and disease, 3(1), 130–140.

World Cancer Research Fund/American Institute for Cancer Research. (2018). Continuous Update Project Report: Diet, Nutrition, Physical Activity, and Colorectal Cancer. Retrieved from https://www.wcrf.org/dietandcancer/colorectal-cancer-report

World Health Organization. (2010). Global recommendations on physical activity for health. Geneva, Switzerland: World Health Organization.

World Health Organization. (2020). Promoting physical activity: A guide to developing and implementing physical activity promotion programmes. Geneva, Switzerland: World Health Organization.

World Health Organization. (n.d.). Who coronavirus (COVID-19) dashboard. World Health Organization. Retrieved December 31, 2022, from https://covid19.who.int/

Post–Covid–19 Indigenous Health

By Annilea Purser

While much social sciences scholarly research has rightfully explored conversations of the COVID-19 pandemic's effect on government practices and responses, placing agency on the government in analyses of the COVID-19 pandemic is necessary to understanding the government's role in both perpetuating and slowing the impacts of the virus. This agency framing has increased importance upon considering certain disparities that have been of focus since the pandemic's first case in Canada in January of 2020, such as the disproportionate effects of COVID-19 on marginalized people; immigrants to Canada who were more likely to live in multigenerational dwellings were subject to increased COVID-19 spread risk; African, Caribbean and Black communities disproportionately account for COVID-19-related deaths; and female-identifying individuals who are more likely to have careers in health and social industries (deemed as essential or front-line services) were faced with increased exposure to COVID-19 (Mishra et. al, 2022, p. 86; Etowa & Hyman, 2020, p.9). Perhaps most prominently in Canada, however, has been COVID-19's effect on Indigenous peoples. While Indigenous peoples across the country generally managed to be affected less than non-Indigenous people during the first wave of the pandemic (January-September 2020) with lower case numbers, the second and third waves of the pandemic (September 2020-August

2021) brought forward significantly increased infection and death rates amongst First Nations, Metis, and Inuit communities, presenting some of the highest COVID-19 rates in the country (House of Commons, 2021, p. 13).

The Government of Canada has continued to frame such a phenomenon as being the product of the "legacy of colonialism," (House of Commons, 2021, p. 15) with the notion of past colonial events continuing to affect Indigenous communities in present day, thus resulting in increased COVID-19 cases and deaths. For example, the Government of Canada's House of Commons (2021) final report on the inquiry into COVID-19 impacts on Indigenous peoples – "COVID-19 and Indigenous Peoples: Moving From Crisis Towards Meaningful Change" – frames said colonial impacts as being those that are "long-standing" or "legacies," (House of Commons, 2021).

This harmful historicization of colonialism during the COVID-19 pandemic easily fits into broader tendencies amongst settler governments. Scholar Patrick Wolfe has made a significant contribution to understanding such a phenomenon, in his 2006 writing Settler colonialism and the elimination of the native. Here, Wolfe convincingly maps genocide and colonialism to ultimately conclude that settler colonialism "is a structure not an event," (Wolfe, 2006, p. 388) considering this land was not only colonized, but the colonizers came to stay. For those unfamiliar with said phenomenon, they may turn to a myriad of examples to better understand the ways in which settler-colonialism acts as a form of ongoing colonization. Perhaps one of the most comprehensive documents that can provide said examples of settler-colonialism being ongoing is the Indian Act of 1876 where various discriminatory practices again Indigenous peoples have been prescribed, including those like the imposition of the band council system – a governance system that works to assimilate traditional Indigenous governance systems – or other assimilatory practices like denying women Indian status based on their gender (ICTINC, 2015).

Thus, the notion of COVID-19 perpetuating settler-colonialism naturally follows the pre-disposed settler-colonial logic.

Thus, we must ask the question, what is the impact of this historicization of colonialism during the COVID-19 pandemic? On a simple level, it is a denial to recognize the notion of Canada as being a settler-state, and on a more complex level of inquiry, it is ignorance of the government's perpetuation of settler-colonial ideals during COVID-19. It is in this context, that a return to investigating the government's role in influencing COVID-19, as an actor with agency, is critical, especially considering the vastly disproportionate effects the pandemic has had on Indigenous communities. Thus, a (re)placing of authority and agency onto the government involves recognizing that, beyond the already-existing forms of health discrimination, government action during the pandemic worked to perpetuate settler-colonialism. The following chapter argues that not only has the COVID-19 pandemic highlighted the ongoing impacts of settler-colonialism on Indigenous peoples, but it has also worked to further the state's settler-colonial project through the prioritization of economic interest over Indigenous livelihoods. To establish this argument, the chapter will: first, briefly establish the dimensions of how Indigenous peoples have been affected by the interplay between settler-colonialism and the pandemic, then explore the dimensions of how Indigenous peoples have been affected by the settler government's economic prioritization during COVID-19; and finally, discuss how conversations may be refocused based on the lessons learned here.

COVID – 19, Settler – Colonialism, & Indigenous Peoples in Canada

Before discussing more in-depth as to how settler-colonialism has been reproduced through the COVID-19 pandemic, it is important to briefly map how Indigenous peoples have been affected more generally. The effects of COVID-19 on Indigenous peoples within Canada can be easily traced far beyond just having greater COVID-related death rates, or being more vulnerable to communicable diseases, with an important role of continued settler-colonialism taking focus (Power et. al, 2020, p. 2738). Said continued effects of settler-colonialism are multidimensional, encompassing a myriad of social and economic issues, all stemming from the impacts of settler-colonialism (Power et. al, 2020). Impacts include those such as threats to cultural continuation with the banning of cultural gatherings which has been harmful to cultural continuity, greater mental illness rates that may be related to isolation or the triggering of pre-exposed stressors/environments, and intersectional issues like increased violence towards Indigenous women and girls (Power, et. al, 2020, p. 2378) To exemplify the complexity of each of these dimensions, one can turn to the specific case of food insecurity – a topic deeply entrenched in colonialism – with geographic (i.e being placed far away from affordable food through the reserve system) and socioeconomic factors resulting in a turn to less expensive and less nutritionally dense foods (Levkoe et. al, 2021, p. 2: Power et. al, 2020, p. 2738). In a pandemic context, this landscape worsened, with cheaper food being less available, food rising in price, and assisted food centres being forced to close due to COVID-19 cases (Levkoe et. al, 2021, p.2). Thus, Indigenous peoples in Canada were not only subject to increased COVID-19 infection and death rates, but also a multitude of other issues, rooted in settler colonialism (Power et. al, 2020).

Extractive Development Displacing Protection and Overriding Consultation

Beyond settler colonialism resulting in disproportionate COVID-19 impacts, governments have utilized the pandemic as a means of reproducing settler colonialism. A telling case of this unsurprisingly falls within the realm of land development, particularly in Western Canada. Scholars like Bernauer and Slowey (2020) in COVID-19, extractive industries, and indigenous communities in Canada: Notes towards a political economy research agenda and Costa (2021) in 'People get what they deserve': necropolitical consultation in the Covid-19 pandemic support such claims by outlining the various dimensions of continued extractive industry development during COVID-19, which is argued here to be a re-production of settler-colonialism. More specifically, and as foregrounded, said settler-colonial reproduction has come to life through the federal and provincial governments' insistence to frame extractive practices as being "economically essential" during the COVID-19 pandemic, while failing to protect Indigenous peoples or engage in proper consultative practices. Since the very beginning of the COVID-19 pandemic, First Nations and Inuit communities alike expressed significant concern over the potential spread of COVID-19 through mechanisms of corporate land development, such as with transient workers travelling into their, often remote, communities to fulfil projects like pipeline installation (Bernauer & Slowey, 2020).

As aforementioned, Indigenous peoples and their subsequent communities face increased social vulnerabilities from long-term colonial effects that result in increased vulnerability and dangers associated with contracting COVID-19 (Power et. al, 2021). This concern is doubly important to consider upon reflecting upon the ageing Indigenous population and the importance of older-aged Indigenous peoples – like elders who are also known as "knowledge keepers" or medicine people – in continuing cultural practices. As Costa (2021) notes,

103

Wet'suwet'en hereditary chiefs wrote a letter amplifying this notion, stating, "Unnecessary death of one language speaker or knowledge keeper would have devastating effects on our families, communities and governance system'" (p. 11). Thus, many Indigenous nations attempted to enforce their own means of protection to defend their important ageing populations and slow the spread of COVID-19 within their communities (House of Commons, 2021, p. 45).

Transient Workers

Despite this, when the federal and provincial governments of Canada were delegating who was to be considered "essential," these transient workers who were at higher risk of transmitting COVID-19 entered these vulnerable Indigenous communities under the guise of continuing economic development through extractive projects. Since the very beginning of the pandemic, Indigenous communities have expressed their concern with such practices (Bernauer & Slowey, 2020, p. 2). A prominent example of this is seen with the Union of BC Indian Chiefs. In an open letter in to members of the Government of Canada in April of 2020, the Union (namely, Grand Chief Stewart Philip, Chief Don Tom, and Kukpi7 Judy Wlson) wrote:

"We urge you to act swiftly to protect the public's health from the heightened risks of COVID-19 transmission posed by ongoing construction of the Coastal GasLink Pipeline Project. Most vulnerable to the spread will be frontline healthcare workers, project workers, and local Indigenous and non-Indigenous communities forced to shoulder the consequences for any disregard for health and safety," (Union of BC Indian Chiefs, 2020).

This letter regarding concerns over Indigenous people's safety during COVID-19 with ongoing extractive projects has been further amplified by already-existing growing worries over general community health in the face of development projects (Bernauer & Slowey, 2020, p. 845). Thus, the Union of BC

104

Indian Chiefs is not alone in expressing such concerns, with similar sentiments arising in uniformly dense extraction jurisdictions, like with the mines of White River First Nation of Beaver Creek, Yukon, and the Mary River iron mine in Nunavut, affecting Inuit communities (Bernauer & Slowey, 2020. p.845).

Consultation

Closely related to the by-passing of keeping Indigenous peoples "safe" during the COVID-19 pandemic with allowing transient workers in vulnerable communities is the topic of lessing Indigenous consultation capacities under pandemic measures – like lockdowns – which has meant lessened consultation rigour for extractive industries, despite already grim procedures. Prior to the pandemic, extractive resource development consultation was found to be a duty of the crown (Boyd & Lorefice, 2018, p. 573). Yet, despite government attempts to perform more significant consultation with Indigenous peoples, significant legal disputes continued to take place over a lack of consultation or insignificant absorption of what was heard during consultation rounds into decision-making, like with the Yahey v. British Columbia case, where consultative efforts failed to bring received comments from Indigenous participants from the Blueberry River First Nation into resource extractive decisions (Terry et. al, 2021). As Andrew Costa (2021) argues in 'People get what they deserve': necropolitical consultation in the Covid-19 pandemic, the current relationship between settler consulting and Indigenous peoples is one that is predicated on "ignoring the conceptual and communal oversights inhering in consultation," (Costa, 2021, p. 11), or in other words, overlooking Indigenous peoples' positionality in consulting work (Costa, 2021, p. 11). Further, consultation hasn't been approached as anything other than checking a box for corporations, with little regulation as to whether feedback is accounted for and concern being placed in filling out paperwork instead of Indigenous livelihoods (Costa, 2021, p. 9).

This precedent of low consultation standards and practices of the Government of Canada and provincial governments like the Government of BC have become even more troubling during the COVID-19 pandemic. As Costa argues, Indigenous consultation under the COVID-19 pandemic has resorted to a necropolitical form of engagement, or in plain language, a form of consultation where governments have aimed to drain Indigenous communities of their resources with little regard to the human effects (Costa, 2021). In practice, this has looked like the continuation of resource extraction consultations during the pandemic that continued with pre-pandemic pacing, thus greatly overwhelming communities already being disproportionately affected (Costa, 2021, p. 9).

Such practices of continuing to allow transient resource extraction industry workers within Indigenous communities and continued consultation – despite their widely, and loudly, expressed concern – speaks to a broader prioritization of settler economic development that contradicts the Government of Canada's framing of "protection" during the COVID-19 pandemic. While the Government of Canada has continued to frame its form of response as a means "to protect Canadians in response to the pandemic," (Government of Canada, 2022) in the face of the aforementioned allowed threats to Indigenous communities, one must question whom the federal government seeks to protect during the pandemic more broadly when considering the fact that they continue to allow resource extraction transient workers into communities.

Justifying the Sidelining of Indigenous Well-being

Although it has yet to be substantively reported on in the COVID-19 context, a frequent resort for justification for the government's continued extractive industry development is in the name of supporting Indigenous communities' economic development, which they often assert will better the well-being of Indigenous communities. However, the reality of economic benefit for Indigenous communities from extractive practices is questionable at best. As eloquently put by Bernauer and Slowey, wealth from development projects continues to "bypass" Indigenous communities, and instead benefit corporations and governments, despite small, recent increases in the "sharing of wealth" (2020, p. 845). Because of the disinformation underlying this notion that Indigenous communities substantially benefit from extractive development, it is also important to look at this issue beyond dollar values and onto the impact of economic framing on a more theoretical scale. As aforementioned, limited scholarly work has been done to explore this issue as it specifically pertains to COVID-19, however, turning to the issue of the economic-benefit framing justification more broadly can offer insight into the bigger picture issue.

Dene scholar Glen Sean Coulthard explores such economic-benefit justification arguments for development in Red Skin White Masks (2014). As Coulthard amplifies, the root of settler-colonialism remains entrenched in "the dispossession of Indigenous peoples of other lands and self-determining authority," (Coulthard, 2014, p. 25) including through land extraction development practices. Coulthard further outlines how this plays out into an economic-benefit framing justification, as with his example of the Delgamuukw v. British Columbia legal case (Coulthard, 2014, p. 41). The Delgamuukw case of 1997 concerns a case wherein First Nations hereditary chiefs filed a land title action to the Supreme Court of British Columbia to receive a mass land area for the purposes of protecting the forest from extractive industry development (Beaudoin, 2019). Ultimately, the court asserted that pursuing "substantial objectives" of the settler government could work as legal justifications for infringing on Indigenous land. Notably, said "substantial objectives" that may justify land infringement all align with the aforementioned economic-benefit framing, including the development of extractive industry. As

Coulthard eloquently states, this consistent prioritization of "economic benefits" – which, as discussed, rarely reach Indigenous populations anyway – points towards the ideas that "colonial powers will only recognize the collective rights and identities of Indigenous peoples insofar as this recognition does not throw into question the background legal, political, and economic framework of the colonial relationship itself," (Coulthard, 2014, p. 41). Such broader theoretical work on economic framing offers important insight into the government's economic-benefit justification of continued extractive development during the COVID-19 pandemic while simultaneously stating that they are "protecting" Indigenous communities. Here, it becomes clear that the motive of settler governments to continue resource extraction development activities despite harming Indigenous populations is not because of the potential economic benefits for Indigenous peoples as they might like to propose, but it is instead about the continued economic gains for the government itself, which they value as being greater than Indigenous livelihoods. In this sense, COVID-19 has worked as a means to justify the furtherment of settler-colonialism.

Lessons Learned & Conclusion

Considering the case study examples of settler governments prioritizing economic values, the argument of settler-colonialism being perpetuated through the COVID-19 pandemic becomesapparent. Importantly, this argument is established in two parts: the first being that settler-colonialism has continued under the pandemic despite life or death circumstances, and the second building on this to assert that the COVID-19 pandemic – through actions like lessing consultative practices – has worked as a form of government justification or scapegoat for even further settler-colonial pursuits than what was viable prior to the pandemic. Such effects become unsurprising upon situating these events within the aforementioned ideas of Patrick Wolfe wherein settler-colonialism is an event, as well as former settler government handlings of pandemic-like conditions. Albeit on a less critical scale, settler governments' prioritization of economic development over the communicable disease health rates in vulnerable communities – like Indigenous communities in Alaska – has been shown many times before (Westwood & Orenstein, 2016, p. 245). Westwood and Orenstein's 2016 study Does resource development increase community sexually transmitted infections? An environmental scan illustrates how other infectious diseases, this time of sexually transmitted nature, have been disproportionately perpetuated by extractive industry transient workers amongst extraction-dense jurisdictions, thus offering a paralleled case for what effects the prioritization of economic development can have on communities with increased vulnerability (Westwood & Orenstein, 2016, p.245). Beyond this case of sexually transmitted diseases is the larger history of the relationship between Indigenous peoples' well-being being sidelined for development, which may be traced back to the first colonial encounters with the mass deaths from communicable diseases like influenza (Hillier et. al, 2020, p. 24).

Therefore, what has unravelled under the COVID-19 pandemic is not a new phenomenon by any means, and will most likely accelerate in the future, unless substantial action to address settler-colonialism is undertaken. In Rehearsals for Living, Black feminist and abolitionist scholar Robyn Maynard highlights the positionality of racialized communities in a climate change world (Maynard, 2022, pp. 58-59). As they note, climate change's developing, and worsening, conditions will mean that we will most likely see a rise in infectious diseases that will undeniably continue to disproportionately affect Indigenous and Black communities, resulting in "bleakness" (Maynard, 2022, pp. 59). Thus, considering the witnessing of settler-colonial perpetuation under COVID-19 and the impending doom of climate change, it is vital that conversations surrounding Indigenous peoples and COVID-19 turn to take action for change.

References

Beaudoin, G. (2017, August 18). Delgamuukw Case. The Canadian Encyclopedia. Retrieved December 30, 2022, from https://www.thecanadianencyclopedia.ca/en/article/delgamuukw-case

Bernauer, W., & Slowey, G. (2020). COVID-19, extractive industries, and indigenous communities in Canada: Notes towards a political economy research agenda. The Extractive Industries and Society, 7(3), 844-846. https://www.ncbi.nlm.nih.gov/pmc/articles/PMC7273156/

Boyd, B., & Lorefice, S. (2018). Understanding consultation and engagement of Indigenous Peoples in resource development: A policy framing approach. Canadian Public Administration, 61(4), 572-595. https://onlinelibrary-wiley-com.proxy.library.brocku.ca/doi/full/10.1111/capa.12301

Costa, A. (2021). 'People get what they deserve': necropolitical consultation in the Covid-19 pandemic. Settler Colonial Studies. https://www-tandfonline-com.proxy.library.brocku.ca/doi/pdf/10.1080/2201473X.2021.2 008100?casa_token=SA_iPogtgJEAAAAA:2W8H92R1KdbAiMKAQE-capN27_exDBOSGmS-C6789LaGAhK_9ihm-LJqf6lyS-Jpm-mM4HR6oJrmT92C8

Coulthard, G. S. (2014). Red Skin, White Masks: Rejecting the Colonial Politics of Recognition. University of Minnesota Press Does resource development increase community sexually transmitted infections? An environmental scan. (2015). The Extractive Industries and Society, 3(1). https://www.researchgate.net/publication/284091458_Does_resource_development_increase_community_sexually_transmitted_infections_An_environmental_scan

Etowa, J., Demeke, J., Abrha, G., Worku, F., Ajiboye, W., Beauchamp, S., Taiwo, I., Pascal, D., & Ghose, B. (2021). Social determinants of the disproportionately higher rates of COVID-19 infection among African Caribbean and Black (ACB) population: A systematic review protocol. J Public Health Res, 11(2), 2274. 10.4081/jphr.2021.2274

Government of Canada. (2022, October 19). Coronavirus (COVID-19) and Indigenous communities. Services aux Autochtones Canada / Indigenous Services Canada. Retrieved December 30, 2022, from https://www.sac-isc.gc.ca/eng/1581964230816/1581964277298

Hillier, S., Chaccour, E., & Al-Shammaa, H. (2020). Indigenous Nationhood in the Age of COVID-19: Reflection on the Evolution of Sovereignty in Settler-Colonial States.

Journal of Indigenous Social Development, 9(3), 23-42. https://journalhosting.ucalgary.ca/index.php/jisd/article/view/70736/54408

House of Commons. (2021). COVID-19 AND INDIGENOUS PEOPLES: FROM CRISIS TOWARD MEANINGFUL CHANGE. https://www.ourcommons.ca/Content/Committee/432/INAN/Reports/RP11143866/inanrp06/inanrp06-e.pdf

ICTINC. (2015, June 2). 21 Things You May Not Have Known About the Indian Act. Indigenous Corporate Training Inc. https://www.ictinc.ca/blog/21-things-you-may-not-have-known-about-the-indian-act

Levkoe, C., McLaughlin, J., & Strutt, C. (2021). Mobilizing Networks and Relationships Through Indigenous Food Sovereignty: The Indigenous Food Circle's Response to the COVID-19 Pandemic in Northwestern Ontario. Frontiers in Communication, 6. https://www.frontiersin.org/articles/10.3389/fcomm.2021.672458/full

Maynard, R., & Simpson, L.B. (2022). Rehearsals for living. Knopf Canada.

Mishra, S., Moloney, G., Ma, H., Yiu, K., Darvin, D., Landsman, D., Kwong, J., & Calzavara, A. (2022). Increasing concentration of COVID-19 by socioeconomic determinants and geography in Toronto, Canada: an observational study. Annals of Epidemiology, 65(1), 84-92. https://www-sciencedirect-com.proxy.library.brocku.ca/science/article/pii/S10472797210 02167?via%3Dihub

Power, T., Wilson, D., Best, O., Brockie, T., Bourque Bearskin, L., Millender, E., & Lowe, J. (2020). COVID-19 and Indigenous Peoples: An imperative for action. Journal of Clinical Nursing, 29(15), 2737-2741. https://www.ncbi.nlm.nih.gov/pmc/articles/PMC7272911/

Union of BC Indian Chiefs. (2020). OPEN LETTER: Coastal GasLink Pipeline Project Must be Halted Due to the COVID-19 Outbreak. Union of British Columbia Indian Chiefs. https://www.ubcic.bc.ca/open_letter_coastal_gaslink_pipeline_project_must_be_halted_d ue_to_the_covid_19_outbreak

Wolfe, P. (2006). Settler colonialism and the elimination of the native. Journal of GenocideResearch, 8, 387-409. https://doi.org/10.1080/14623520601056240